Holt McDougal
Mathematics

Course 3
Chapter 10 Resource Book

Contents

Holt McDougal Mathematics

Description of Contents

Family Involvement Pages

The Chapter Resource Book includes a set of family involvement pages for each section. The family involvement pages consist of the following items:

- **Family Letter**, a two-page letter describing the math that the student will study in each section. A list of vocabulary words from the lessons is included, and some worked-out examples are provided.
- **At-Home Practice**, a one-page worksheet with problems drawn from the content described in the Family Letter. Answers are provided on the page so that the student and his or her family can check this work at home.
- **Family Fun**, a one-page activity sheet that the student and his or her family can work on together.

Practice A, B, and C

There are three practice worksheets for every lesson. All of these reinforce the content of the lesson. Practice B is shown in the Teacher's Edition and is appropriate for the on-level student. It is also available as a workbook (Homework & Practice Workbook).

Practice A is easier than Practice B but still practices the content of the lesson. Practice C is more challenging than Practice B.

Review for Mastery

The Review for Mastery worksheet (one per lesson) provides an alternate way to teach or review the main concepts of the lesson. This worksheet is one or two pages long and is shown in the Teacher's Edition.

Challenge

The Challenge worksheet (one per lesson) enhances critical thinking skills and extends the lesson. This worksheet is shown in the Teacher's Edition.

Problem Solving

The Problem Solving worksheet (one per lesson) provides practice in problem solving and opportunities for real-world applications and for interdisciplinary connections. There are both multiple choice and short response problems. This worksheet is shown in the Teacher's Edition.

Reading Strategies

The Reading Strategies worksheet (one per lesson) provides tools to help the student master math vocabulary or symbols.

Puzzles, Twisters & Teasers

The Puzzles, Twisters & Teasers worksheet (one per lesson) provides fun practice while reinforcing the content of the lesson.

Holt McDougal Mathematics

Family Letter

10A Experimental and Theoretical Probability

Dear Family,

The student will be learning the necessary terminology for understanding and applying the concepts of **probability**.

Here is how the student will learn to find the probabilities of **outcomes** in a **sample space**.

A bank teller's cash drawer contains 4 types of bills; $1, $5, $10, and $20. The table gives the probability of randomly selecting each type of bill from the drawer.

Bill	$1	$5	$10	$20
Probability	0.36	0.30	0.24	0.10

A. What is the probability of selecting a $1 or a $5?
$P(\$1 \text{ or } \$5) = P(\$1) + P(\$5) = 0.36 + 0.30 = 0.66$

B. What is the probability of not selecting a $1?
Since the probabilities must add up to 1, you can subtract.

$P(\text{not selecting } \$1) = 1 - P(\text{selecting } \$1) = 1 - 0.36 = 0.64$

The student will be introduced to **experimental probability**.

A large aquarium contains goldfish, black fish, and red fish. Julie recently has seen 5 goldfish, 7 black fish, and 3 red fish. What is the probability that the next fish she sees will be a goldfish?

Outcomes	Gold	Black	Red
Observations	5	7	3

$$\frac{\text{number of goldfish}}{\text{total number of fish}} = \frac{5}{5+7+3} = \frac{5}{15} = \frac{1}{3}$$

The probability that Julie sees a goldfish next is $\frac{1}{3}$ or 33%.

You can find the probability of an event without performing an experiment by calculating its **theoretical probability**.

An experiment consists of rolling a number cube and flipping a coin. There are 12 possible outcomes: H1, H2, H3, H4, H5, H6, T1, T2, T3, T4, T5, and T6. What is the probability of getting a heads and rolling an odd number?

There are 3 outcomes in the event of getting a heads and rolling an odd number: H1, H3, and H5.

Vocabulary

These are the math words we are learning:

complement all of the outcomes not in the sample space

compound event is made up of two or more separate events

dependent events events for which the outcome of the first event affects the outcome of the second event

disjoint events events that cannot occur in the same trial of an experiment

equally likely outcomes that have the same probability

event a set of one or more outcomes of an experiment

experiment an activity in which the results are observed

experimental probability the ratio of the number of times an event occurs to the total number of trials

geometric probability is a theoretical probability based on the ratio of geometric lengths, areas, or volumes

fair when all outcomes of an experiment are equally likely

Holt McDougal Mathematics

$P(\text{heads and odd}) = \dfrac{3}{12} = \dfrac{1}{4}.$

To find the probability that two **independent events** occur, the student will multiply the probability that the first event happens by the probability that the second event happens.

Joey is holding two hats. Each one contains 26 pieces of paper on which each letter of the alphabet is written. If you choose a letter from each hat, what is the probability that you choose a vowel from each?

$P(\text{vowel from 1}^{st}\text{ hat}) = \dfrac{5 \text{ vowels}}{26 \text{ letters}}$

$P(\text{vowel from 2}^{nd}\text{ hat}) = \dfrac{5 \text{ vowels}}{26 \text{ letters}}$

$P(\text{vowel and vowel}) = \dfrac{5}{26} \cdot \dfrac{5}{26} = \dfrac{25}{676}$

To find the probability that two **dependent events** occur, the student will multiply the probability that the first event happens by the probability that the second event happens. The student must keep in mind that the first event will affect the probability of the second event.

Joey is now holding one hat containing 26 pieces of paper on which each letter of the alphabet is written. If you choose 2 letters from the hat, what is the probability that both are vowels?

$P(\text{vowel from 1}^{st}\text{ choice}) = \dfrac{5 \text{ vowels}}{26 \text{ letters}}$

$P(\text{vowel from 2}^{nd}\text{ choice}) = \dfrac{4 \text{ vowels left}}{25 \text{ letters left}}$

$P(\text{vowel and vowel}) = \dfrac{5}{26} \cdot \dfrac{4}{25} = \dfrac{2}{65}$

Help the student understand probability by relating it to everyday events.

Sincerely,

---

independent events the occurrence of one event has no effect on the probability that a second event will occur

mutually exclusive events that cannot occur in the same trial of an experiment

outcome a possible result of an experiment

probability a number from 0 to 1 that tells how likely an event is to happen

sample space the set of all possible outcomes of an experiment

simulation a model of a real situation

theoretical probability the ratio of the number of favorable outcomes to the number of possible outcomes

trial the act of trying, testing, or putting to the proof

Holt McDougal Mathematics

CHAPTER 10

At-Home Practice

10A Experimental and Theoretical Probability

Give the probability of each outcome or event.

1. rolling a 2 on a number cube ____________________

2. rolling an odd number on a number cube ____________________

An experiment consists of reaching into a bag, pulling out a handful of mixed coins, and counting the number of pennies. The table gives the probability of each outcome.

Outcome	0	1	2	3
Probability	0.14	0.36	0.41	0.09

3. What is the probability of pulling out exactly 1 penny? ____________________

4. What is the probability of pulling out at least 1 penny? ____________________

5. What is the probability of not pulling out a penny? ____________________

An experiment consists of rolling a fair number cube and flipping a coin. Find each probability.

6. rolling a 5 ____________________

7. rolling a number less than 5 and getting a heads ____________________

A bag contains 7 red marbles, 2 blue marbles, and 1 green marble. Find each probability.

8. Find the probability of drawing a red marble, replacing it, and drawing another red marble. ____________________

9. Find the probability of drawing two red marbles if you do not replace the first marble. ____________________

Answers: 1. $\frac{1}{6}$ 2. $\frac{1}{2}$ 3. 0.36 4. 0.86 5. 0.14 6. $\frac{1}{6}$ 7. $\frac{1}{3}$ 8. $\frac{49}{100}$ 9. $\frac{7}{15}$

Holt McDougal Mathematics

<table>
<tr><td>CHAPTER
10</td><td>Family Fun
Probability Word Search</td></tr>
</table>

Directions

- Fill in the blanks below with the appropriate vocabulary words.

- Locate each of the words in the word search puzzle and circle it.

- The player who finds all nine words first wins the game.

1. An ____________ is a set of one or more outcomes.

2. The ____________ of an event is a number from 0 to 1 that tells how likely the event is to happen.

3. An ____________ is an activity in which results are observed.

4. A result of an experiment is an ____________.

5. The ____________ is the set of all possible outcomes to an experiment.

6. Each observation of a test is called a ____________.

7. ____________ probability is the ratio of the number of times an event occurs to the number of trails.

8. A ____________ is a model of a real situation.

9. ____________ numbers are a set of numbers in which there is no pattern.

```
E C A P S E L P M A S X W O E
S X K I E E F H L E A T B U J
I M P O S S I B L E Z I N T L
M H A E H L O E J T M I G C B
U S B M R H U P W S A F H O M
L H Q D A I M G L T Y H X M U
A L Z T N E M I R E P X E E L
T B J K D O G E V E N T J N A
I L M H O E C A N E P R D L L
O P S I M E T K X T N I D H M
N T E L W O D A N T A A K Q O
A E H M P R O B A B I L I T Y
```

Answers: 1. event **2.** probability **3.** experiment **4.** outcome **5.** sample space **6.** trial
7. Experimental **8.** simulation **9.** Random

Holt McDougal Mathematics

<table>
<tr><td>LESSON
10-1</td><td><h1>Practice A</h1>Probability</td></tr>
</table>

1. Meteorology is the study of the atmosphere, natural phenomenon, atmospheric conditions, weather and climate. A meteorologist forecasts and reports the weather. A meteorologist forecasts a 70% chance of rain. What is the probability of each outcome?

Outcome	Rain	No rain
Probability		

Use the spinner to determine the probability of each outcome.

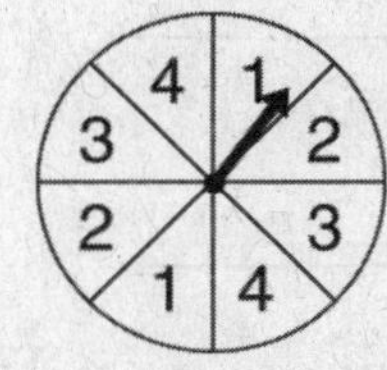

2. $P(1)$

3. $P(4)$

4. $P(\text{even number})$

5. $P(5)$

6. $P(\text{odd number})$

7. $P(\text{an integer})$

8. $P(1 \text{ or } 2)$

9. $P(\text{number} > 1)$

10. $P(\text{a whole number})$

11. Mrs. Silverstein has 14 boys in her class of 25 students. She must select one student at random to serve as the class moderator. What is the probability that she will choose a boy? What is the probability that she will choose a girl?

12. When tossing a regular coin, what is the probability of it landing on heads?

Practice B
Probability

These are the results of the last math test. The teacher determines that anyone with a grade of more than 70 passed the test. Give the probability for the indicated grade.

Grade	65	70	80	90	100
# of Students	5	3	12	10	2

1. $P(70)$
2. $P(100)$
3. $P(80)$
4. $P(\text{passing})$

5. $P(\text{grade} > 80)$
6. $P(60)$
7. $P(\text{failing})$
8. $P(\text{grade} \leq 80)$

A bowling game consists of rolling a ball and knocking up to 5 pins down. The number of pins knocked down are then counted. The table gives the probability of each outcome.

Number of Pins Down	0	1	2	3	4	5
Probability	0.175	0.189	0.264	0.205	0.132	0.035

9. What is the probability of knocking down all 5 pins?

10. What is the probability of knocking down no pins?

11. What is the probability of knocking down at most 2 pins?

12. Four friends are playing a game. Jordan has a 45% chance of winning. Diane is twice as likely to win than Carlos, while Stacey has a 25% chance. Create a table of probabilities for the sample space.

Outcome				
Probability				

Holt McDougal Mathematics

LESSON 10-1 · Practice C
Probability

Demographers often use statistics to predict and explain future changes in populations in many areas including housing, education, life events, and unemployment. A demographer developed this chart to illustrate the cause of death in his community of 50,000 people.

Give the probability for each outcome.

Outcome	Heart	Cancer	Accident	Respiratory	Other
Probability	0.35	0.28	0.16	0.13	0.08

1. P(death from cancer)   2. P(death from accident)   3. P(death from heart)

_______________   _______________   _______________

4. P(non-accidental death)   5. P(death from other)   6. P(death from heart or cancer)

_______________   _______________   _______________

Use the spinner to determine the probability of each outcome.

7. P(white 1)   8. P(dots 2)   9. P(lines even)   10. P(dots 1)

_______________   _______________   _______________   _______________

11. P(white odd)   12. P(dots integer)   13. P(odd)   14. P(white or 2)

_______________   _______________   _______________   _______________

15. There are six teams competing to collect the most food for the food bank. Team B has a 30% chance of winning. Teams A, C, D, and E all have the same chance of winning. Team F is one-third as likely to win as Team B. Create a table of probabilities for the sample space.

Outcome						
Probability						

LESSON 10-1 — Review for Mastery
Probability

The **probability** that something will happen is how often you can expect that **event** to occur. This depends upon how many outcomes are possible, the **sample space.**

In the spinner shown, the circle is divided into four equal parts. There are 4 possible outcomes. So, in a single spin:

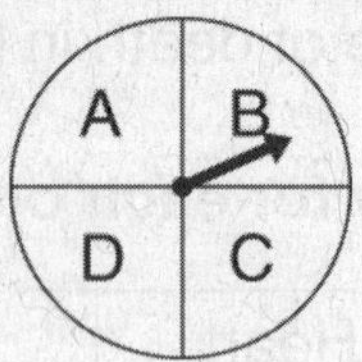

$$P(A) = P(B) = P(C) = P(D) = 25\% = \frac{1}{4}$$

Complete to give the probability for each event.

	1. A fair coin is tossed.	2. A number cube is rolled.
List all the possible outcomes.	____________ ____________	____________ ____________
How many outcomes in sample space?	____________	____________
Find the probability of the event shown.	$P(\text{heads}) =$ ________	$P(5) =$ ________

• The **complement** of an event is all of the outcomes not in the sample space.
• The sum of the probabilities of an event and its complement is 1.
  If the probability of *snow* is 30%, then the probability of *no snow* is 70%.
  $$P(\text{snow}) + P(\text{no snow}) = 1$$

3. The probability of a coin landing on heads is 50%, so the probability of landing on tails is:

____

4. If the probability of choosing a blue marble in a bag is 0.72, then the probability of not choosing a blue marble is:

____

5. If the probability of selecting a senior for a committee is 60%, then the probability of not selecting a senior is:

____

6. If the probability of choosing a red ball from a certain box is 0.35, then the probability of not choosing a red ball is:

____

Holt McDougal Mathematics

Review for Mastery
LESSON 10-1

Probability (continued)

To find the probability that an event will occur, add the probabilities of all the outcomes included in the event.

This bar graph shows the midterm grades of the 30 students in Ms. Lin's class.

What is the probability that Susan has a grade of C or higher?

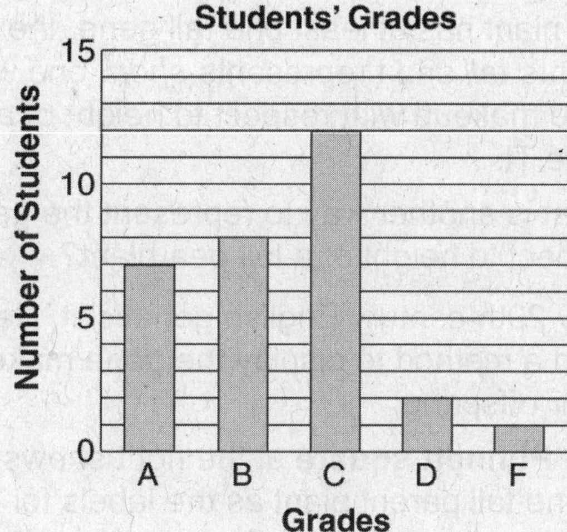

The event "a grade of C or higher" consists of the outcomes C, B, A.

$P(\text{C or higher}) = P(\text{C}) + P(\text{B}) + P(\text{A})$

$$= \frac{12}{30} + \frac{8}{30} + \frac{7}{30} = \frac{12 + 8 + 7}{30} = \frac{27}{30} = \frac{9}{10} = 90\%$$

So, the probability that Susan's midterm grade is C or higher is $\frac{9}{10}$ or 90%.

Use the bar graph above to find each probability.

7. B or higher is honor roll. What is the probability that Ken made the honor roll?

$P(\text{B or higher}) = P(\text{B}) + \underline{\quad\quad}$

$$= \frac{\quad}{30} + \frac{\quad}{30}$$

$$= \frac{\quad}{30} = \underline{\quad\quad}$$

So, the probability that Ken made the

honor roll is: _______ or _______ %.

8. In this class, D is a failing grade. What is the probability that Tom failed?

$P(\text{D or lower}) = P(\text{D}) + \underline{\quad\quad}$

$$= \frac{\quad}{30} + \frac{\quad}{30}$$

$$= \frac{\quad}{30} = \underline{\quad\quad}$$

So, the probability that Tom

failed is: _______ or _______ %.

Holt McDougal Mathematics

Name _______________________________   Date ________________   Class ____________

LESSON 10-1

Challenge

Why We Look Like Our Parents

Each parent carries two genes with respect to a specific trait and each passes one of these genes on to an offspring who then also has two genes for that trait.

In pea plants, a tall gene is dominant over a short gene. So, if a pea plant has at least one tall gene, the plant is tall. If T represents *tall* and t represents *short*, one way to represent the gene makeup with respect to height of a tall pea plant would be Tt.

1. What is another way to represent the gene makeup with respect to height of a tall pea plant?

An early 20th-century English geneticist, Reginald Punnett, invented a method to display the gene makeup of parents and their offspring.

2. The **Punnett square** at the right shows the gene makeup of one tall parent plant as the labels for the columns. Insert your result from Question 1 for the other tall parent plant as the row labels.

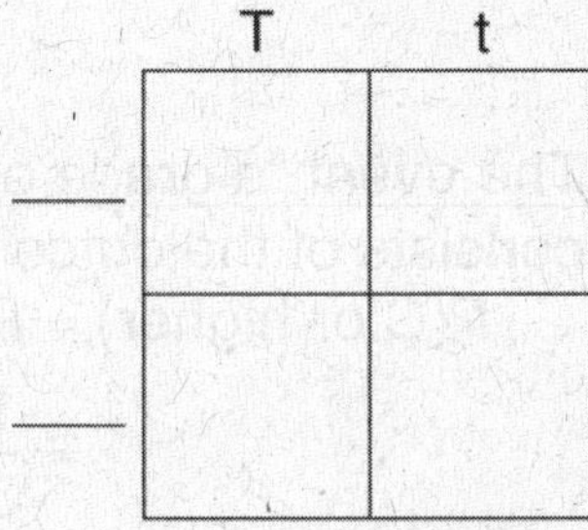

3. a. The label for each column has been inserted in each box of its column, as the first gene of the offspring plant. Insert your labels for each row as the second gene of each offspring plant.

   b. According to the two genes now in each of the boxes for the new offspring plants, tell if the new plant will be tall or short.

   c. What is the probability that an offspring of these tall parent plants will be tall?

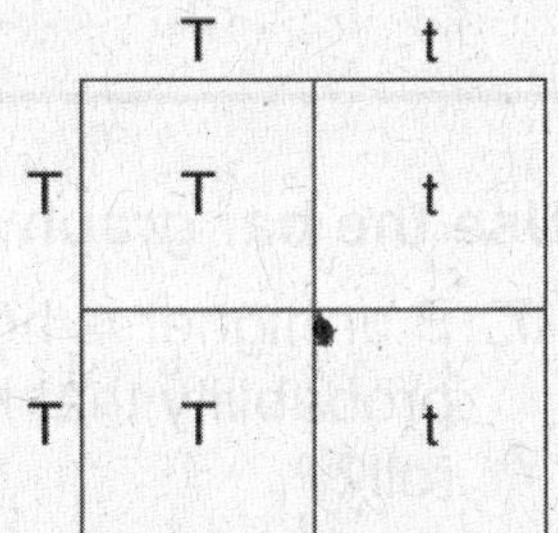

4. Suppose the gene makeup for both tall parent pea plants is Tt.

   a. Complete a Punnett square to display the gene makeup of the offspring.

   b. What is the probability that an offspring of these tall parent plants will be tall?

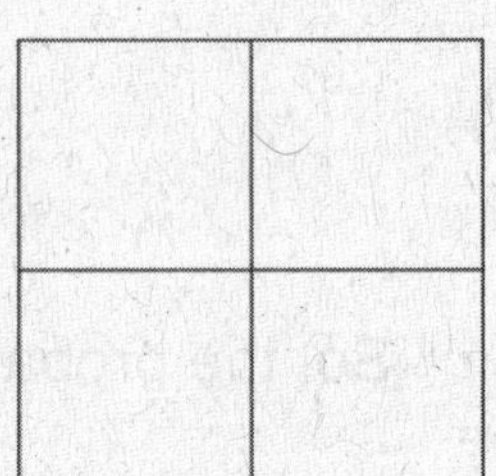

Holt McDougal Mathematics

LESSON 10-1 — Problem Solving
Probability

Write the correct answer.

1. To get people to buy more of their product, a company advertises that in selected boxes of their popsicles is a super hero trading card. There is a $\frac{1}{4}$ chance of getting a trading card in a box. What is the probability that there will not be a trading card in the box of popsicles that you buy?

____

2. The probability of winning a lucky wheel television game show in which 6 preselected numbers are spun on a wheel numbered 1–49 is $\frac{1}{13,983,816}$ or 0.000007151%. What is the probability that you will not win the game show?

____

Based on world statistics, the probability of identical twins is 0.004, while the probability of fraternal twins is 0.023.

3. What is the probability that a person chosen at random from the world will be a twin?

____

4. What is the probability that a person chosen at random from the world will not be a twin?

____

Use the table below that shows the probability of multiple births by country. Choose the letter for the best answer.

5. In which country is it most likely to have multiple births?

   A Japan

   B United States

   C Sweden

   D Switzerland

6. In which country is it least likely to have multiple births?

   F Japan

   G United States

   H Sweden

   J Switzerland

7. In which two countries are multiple births equally likely?

   A United Kingdom, Canada

   B Canada, Switzerland

   C Sweden, United Kingdom

   D Japan, United States

Probability of Multiple Births

Country	Probability
Canada	0.012
Japan	0.008
United Kingdom	0.014
United States	0.029
Sweden	0.014
Switzerland	0.013

Holt McDougal Mathematics

LESSON 10-1	**Reading Strategies**
	Focus On Vocabulary

Probability is the chance that something will happen.

An **event** has an outcome that can be stated using probability. The probability of something happening is written **P(event)**.

The probability of an event happening is described by a number between 0 and 1, or as a percent between 0% and 100%.

The probability of an event *not* to happen is the event's **complement**.

Answer each question.

1. What name is given to the chance of something happening?

   ____

2. How is the probability of an event written?

   ____

3. What number or percent is used to show the sum of the probability of an event and its complement?

   ____

An **experiment** is an activity where results are observed. Flipping a coin and tossing a number cube are both experiments.

In an experiment, each observation (such as one coin toss) is called a **trial**. The result of the trial is called an **outcome** (such as a coin landing on heads).

Use the terms above to answer the following questions.

4. What do you call an activity in which results are observed?

   ____

5. What is one observation in an experiment called?

   ____

6. What is an outcome?

   ____

Holt McDougal Mathematics

LESSON 10-1

Puzzles, Twisters & Teasers
Probability

Solve the crossword puzzle.

Across

3. The _____ of an event is all of the outcomes not in the sample space.

5. The probabilities of all the outcomes in the sample space add up to _____.

7. An _____ is any set of one or more outcomes.

8. Each observation in an experiment is called a _____.

Down

1. Each result of an experiment is called an _____.

2. The _____ of an event is a number that tells how likely the event is to happen.

4. An _____ is an activity in which results are observed.

6. The set of all possible outcomes of an experiment is the _____ space.

Holt McDougal Mathematics

Practice A
Experimental Probability

The results of an unbiased survey show the favorite instruments of 8th graders. Estimate the probability of each.

Result	Piano	Drums	Trombone	Flute	Violin	Clarinet
Number	1	4	42	38	12	3

1. a student chooses clarinet

   ____

2. a student chooses drums

   ____

3. a student chooses flute

   ____

4. a student chooses piano

   ____

5. a student chooses trombone

   ____

6. a student chooses violin

   ____

A can contains color chips in 5 different colors. Thomas took a sample from the can and counted the colors. His results are in the table below.

Color	Blue	Pink	Black	White	Green
Number	10	5	20	30	15

7. Use the table to compare the probability that Thomas chooses a pink color chip to the probability that he chooses a white color chip.

   ____

8. Use the table to compare the probability that Thomas chooses a green color chip to the probability that he chooses a blue color chip.

   ____

9. Cheryl surveyed 30 students who ride the bus to school, 8 who walk, 9 who ride bicycles, and 3 who ride in cars. Estimate the probability that the next student Cheryl surveys will walk to school. ____________________

 Holt McDougal Mathematics

Practice B
Experimental Probability

1. A movie theater sells popcorn in small, medium, large and jumbo sizes. The customers of the first show purchase 4 small, 20 medium, 40 large, and 16 jumbo containers of popcorn. Estimate the probability of the purchase of a medium container of popcorn.

   ____

2. Melissa was able to make 6 free throws out of her last 10 attempts. Estimate the probability that she will hit the next free throw.

   ____

3. The table shows the different bus routes available to the residents of a city. Estimate the probability that a resident will take Route A.

Bus Route	Route A	Route B	Route C	Route D
Residents	105	61	434	30

   ____

4. Which DJ has the highest probability that a song played will be classical?

DJ Songlists		
DJ	Classical Songs	Number of Songs
Thomas	46	542
Guy	21	134
Paul	132	2010

   ____

5. Two in every 3 gift bags will have an extra prize. Simulate by using a number cube, and estimate the probability that 6 bags in 9 will have an extra prize.

   ____

Holt McDougal Mathematics

LESSON 10-2

Practice C

Experimental Probability

1. The developer of a Web page wants to track the number of hits to each link of the Web page. An automatic counter records the following hits in one week: home, 60 hits; FAQ, 20 hits; employment opportunities, 15 hits; products, 50 hits; order status, 30 hits; and contact information, 25 hits. Estimate the probability of a hit on the contact information page.

____

The table shows how many times a song on a CD was played at a party.

Track	1	2	3	4	5	6	7	8	9	10	11	12
Frequency	2	4	3	1	2	4	2	3	5	2	1	4

Estimate the probability for each of the following.

2. $P(\text{track } 2)$

3. $P(\text{track } 4)$

4. $P(\text{track } 5)$

____

5. $P(\text{track } 8)$

6. $P(\text{track } 9)$

7. $P(\text{track } 13)$

____

8. Use the table to compare the probability that Track 10 was played to the probability that Track 6 was played.

____

9. About 1 in 6 students will get a brain teaser correct. Simulate by using a number cube, and estimate the probability that 2 students out of 6 will get the correct answer.

____

LESSON 10-2	**Review for Mastery**
	Experimental Probability

A machine is filling boxes of apples by choosing 50 apples at random from a selection of six types of apples. An inspector records the results for one filled box in the table below.

Type	Pink Lady	Red Delicious	Granny Smith	Golden Delicious	Fuji	MacIntosh
Number	8	12	6	4	15	5

The inspector then expands the table to find the experimental probability.

$$\text{probability} = \frac{\text{number of type of apple}}{\text{total number of apples}}$$

Type	Pink Lady	Red Delicious	Granny Smith	Golden Delicious	Fuji	MacIntosh
Experimental Probability (ratio)	$\frac{8}{50}$, or $\frac{4}{25}$	$\frac{12}{50}$, or $\frac{6}{25}$	$\frac{6}{50}$, or $\frac{3}{25}$	$\frac{4}{50}$, or $\frac{2}{25}$	$\frac{15}{50}$, or $\frac{3}{10}$	$\frac{5}{50}$, or $\frac{1}{10}$
Experimental Probability (percent)	16%	24%	12%	8%	30%	10%

Find each sum for the apple experiment.

1. The sum of the experimental probability ratios.

$$\text{probability} = \frac{8}{50} + \frac{12}{50} + \frac{6}{50} + \frac{4}{50} + \frac{15}{50} + \frac{5}{50} = \frac{\ \ \ }{50} \text{ or } \_\_\_\_$$

2. The sum of the experimental probability percents.

$$\text{probability} = 16\% + 24\% + 12\% + 8\% + 30\% + 10\% = \_\_\_\_\% \text{ or } \_\_\_\_$$

Complete the table to find the experimental probability.

3. Five types of seed are inserted at random in a pre-seeded strip ready for planting.

Type	Marigold	Impatiens	Snapdragon	Daisy	Petunia
Number	40	100	80	60	120
Experimental Probability (ratio)	$\frac{\ \ \ }{400}$, or $\_\_\_$	$\frac{\ \ \ }{400}$, or $\_\_\_$	$\frac{\ \ \ }{400}$, or $\_\_\_$	$\frac{\ \ \ }{400}$, or $\_\_\_$	$\frac{\ \ \ }{400}$, or $\_\_\_$
Experimental Probability (percent)					

    Holt McDougal Mathematics

Name _______________________________ Date _____________ Class _____________

Challenge
Tossing and Spinning

The more times you repeat an experiment, the closer the
experimental probability and the theoretical probability become.

Toss a penny 200 times.

	Heads	Tails

1. Record your results in the table.

2. What is the theoretical probability of:

   getting heads? _____________ getting tails? _____________

3. What is your experimental probability of:

   getting heads? _____________ getting tails? _____________

4. How close are your experimental probabilities to the theoretical probabilities?

   ____

Spin a penny 200 times.

	Heads	Tails

5. Record your results in the table.

6. What is your experimental probability of:

   getting heads? _____________ getting tails? _____________

7. Compare your experimental probabilities for tossing the penny and spinning the
   penny. Are they close? Explain.

   ____

Roll a number cube 200 times.

1	2	3	4	5	6

8. Record your results in the table.

9. What is the theoretical probability of:

   getting a 1? ______ a 2? ______ a 3? ______ a 4? ______ a 5? ______ a 6? ______

10. What is your experimental probability of getting:

   a 1? ______ a 2? ______ a 3? ______ a 4? ______ a 5? ______ a 6? ______

11. How close are your experimental probabilities to the theoretical probabilities?

Holt McDougal Mathematics

<table>
<tr><td>LESSON
10-2</td></tr>
</table>

Problem Solving
Experimental Probability

Use the table below. Round to the nearest percent. Write the correct answer.

1. Estimate the probability of sunshine in Buffalo, NY.

   ____

2. Estimate the probability of sunshine in Fort Wayne, IN.

   ____

Average Number of Days of Sunshine Per Year for Selected Cities

City	Number of Days
Buffalo, NY	175
Fort Wayne, IN	215
Miami, FL	256
Raleigh, NC	212
Richmond, VA	230

3. Estimate the probability of sunshine in Miami, FL.

   ____

4. Estimate the probability that it will not be sunny in Raleigh, NC.

   ____

5. Estimate the probability that it will not be sunny in Miami, FL.

   ____

6. Estimate the probability of sunshine in Richmond, VA.

   ____

Use the table below that shows the number of deaths and injuries caused by lightning strikes. Choose the letter for the best answer.

7. Estimate the probability of being injured by a lightning strike in New York.

   A 0.0000007%   C 0.00007%

   B 0.0000002%   D 0.000002%

States with Most Lightning Deaths

State	Average deaths per year	Average injuries per year	Population
Florida	9.6	32.7	15,982,378
North Carolina	4.6	12.9	8,049,313
Texas	4.6	9.3	20,851,820
New York	3.6	12.5	18,976,457
Tennessee	3.4	9.7	5,689,283

8. Estimate the probability of being killed by lightning in North Carolina.

   F 0.0000006%   H 0.00002%

   G 0.00006%   J 0.000002%

9. Estimate the probability of being struck by lightning in Florida.

   A 0.00006%

   B 0.00026%

   C 0.0000026%

   D 0.0006%

10. In which two states do you have the highest probability of being struck by lightning?

   F Florida, North Carolina

   G Florida, Tennessee

   H Texas, New York

   J North Carolina, Tennessee

     Holt McDougal Mathematics

Reading Strategies

LESSON 10-2

Make Predictions

Experimental probability is a statement of the results of a number of trials.

$$\text{Probability} = \frac{\text{number of times an event happens}}{\text{total number of trials}}$$

When you spin this spinner, it could land on the section with dots, the striped section, or the white section.

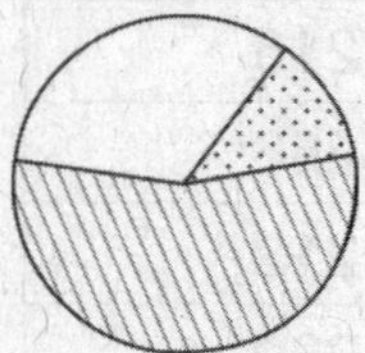

Use the spinner to answer the following questions.

1. Predict which section the spinner will land on most often. Why?

   ____

2. Predict which section the spinner will land on least often. Why?

   ____

The actual outcome of an experiment may or may not match your predictions. This chart shows the outcomes of 500 trials.

Outcome	Striped	Spotted	White
Spins	163	152	185

Answer the following questions.

3. How many times did the spinner land on the striped section? _______________

4. Did *P*(striped) match your prediction for the outcome of the experiment? Explain.

   ____

   ____

5. Did *P*(spotted) match your prediction for the outcome of the experiment? Explain.

   ____

   ____

Puzzles, Twisters & Teasers
Probable Problems

LESSON 10-2

Estimate the probability of drawing a yellow marble in each situation below. Use the letters to answer the riddle.

1. probability: _______% **E**

Outcome	Blue	Yellow	Green	Black
Draws	217	201	295	287

2. probability: _______% **W**

Outcome	Blue	Red	Yellow
Draws	448	267	285

3. probability: _______% **V**

Outcome	Blue	Yellow	Green	White
Draws	33	13	46	8

4. probability: _______% **Y**

Outcome	Blue	Yellow	Green
Draws	330	233	437

5. probability: _______% **C**

Outcome	Blue	Yellow	Green	White	Red
Draws	20	17	32	18	13

6. probability: _______% **L**

Outcome	Blue	Yellow
Draws	65	35

7. probability: _______% **H**

Outcome	Blue	Red	Yellow	White	Green
Draws	30	18	18	21	13

8. probability: _______% **D**

Outcome	Blue	Yellow	Green	Black	White
Draws	17	11	25	28	19

9. probability: _______% **A**

Outcome	Blue	Yellow	Green	White	Red	Black
Draws	65	35	235	180	369	116

10. probability: _______% **N**

Outcome	Blue	Yellow	Green	Black	White	Brown
Draws	7	21	15	18	19	20

What do Christmas and a cat on the beach have in common?

They both ____ ____ ____ ____   **SA** ____ ____ ____

      18   3.5   13   20.1   21   11   23.3

    ____ ____ **A** ____ **S**

    17   35   28.5

Holt McDougal Mathematics

LESSON 10-3 Practice A
Theoretical Probability

An experiment consists of tossing two coins.

1. List all the possible outcomes. _________________

2. What is the probability of tossing a head and a tail? _________________

3. What is the probability of the outcomes being the same? _________________

An experiment consists of rolling a fair number cube.
Find the probability of each event.

4. $P(6)$

5. $P(1)$

6. $P(\text{odd number})$

7. $P(>4)$

Find the probability of each event using two number cubes.

8. $P(\text{rolling two 5s})$

9. $P(\text{total shown} = 4)$

10. $P(\text{total shown} = 2)$

11. $P(\text{total shown} < 4)$

12. $P(\text{total shown} > 11)$

13. $P(\text{rolling two even numbers})$

14. Find the probability that a point chosen randomly inside the square is within the circle. Round to the nearest hundredth.

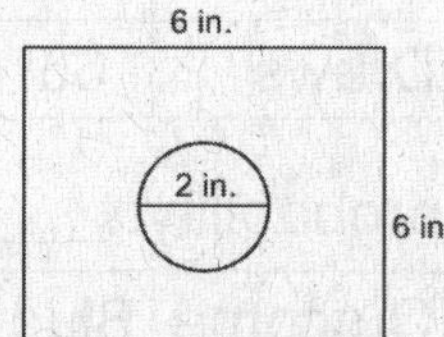

15. In a game two fair number cubes are rolled. To make the first move, you need to roll an even total. What is the probability of rolling an even total? _________________

Holt McDougal Mathematics

Practice B

LESSON 10-3

Theoretical Probability

**An experiment consists of rolling one fair number cube.
Find the probability of each event.**

1. $P(3)$

2. $P(7)$

3. $P(1$ or $4)$

4. $P(\text{not } 5)$

5. $P(< 5)$

6. $P(> 4)$

7. $P(2$ or odd$)$

8. $P(\leq 3)$

**An experiment consists of rolling two fair number cubes.
Find the probability of each event.**

9. $P(\text{total shown} = 3)$

10. $P(\text{total shown} = 7)$

11. $P(\text{total shown} = 9)$

12. $P(\text{total shown} = 2)$

13. $P(\text{total shown} = 4)$

14. $P(\text{total shown} = 13)$

15. $P(\text{total shown} > 8)$

16. $P(\text{total shown} \leq 12)$

17. $P(\text{total shown} < 7)$

18. Find the probability that a point chosen randomly inside the triangle is within the square. Round to the nearest hundredth.

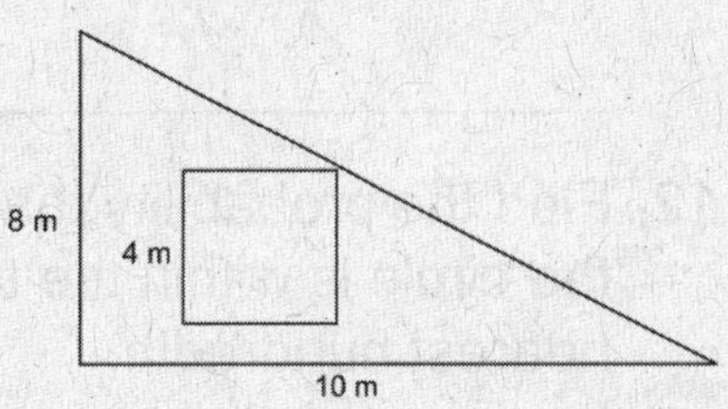

19. In a game two fair number cubes are rolled. To make the first move, you need to roll a total of 6, 7, or 8. What is the probability that you will be able to make the first move? _____________

Holt McDougal Mathematics

LESSON 10-3 Practice C
Theoretical Probability

An experiment consists of rolling two fair number cubes. Find the probability of each event.

1. P(total shown = 5)

2. P(total shown > 3)

3. P(total shown > 10)

4. P(total shown < 12)

5. P(total shown ≥ 7)

6. P(total shown ≤ 4)

Three separate jars each contain 2 different color marbles. Jar A has a red and a blue marble. Jar B has a red and a green marble. Jar C has a purple and a white marble. One marble is drawn from each jar. The table shows a sample space with all outcomes equally likely. Find each probability.

Jar A	Jar B	Jar C	Outcome
R	R	P	RRP
R	R	W	RRW
R	G	P	RGP
R	G	W	RGW
B	R	P	BRP
B	R	W	BRW
B	G	P	BGP
B	G	W	BGW

7. P(RRP)

8. P(BGW)

9. P(2 red with another color)

10. P(a green with two other colors)

11. P(1 white or 1 purple)

12. Find the probability that a point chosen randomly inside the circle is within the triangle. Round to the nearest hundredth.

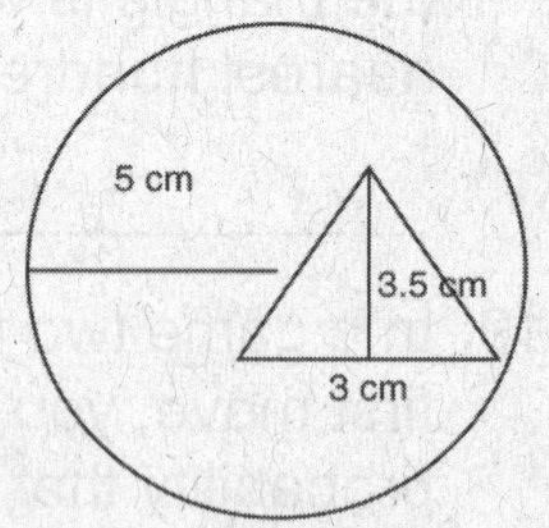

13. In a game two fair number cubes are rolled. To make the first move, you need to roll a total of 7, 8, or 9. What is the probability that you will be able to make the first move? _______________

 Holt McDougal Mathematics

LESSON 10-3
Review for Mastery
Theoretical Probability

The sample space for a fair coin has 2 possible outcomes: heads or tails. Both possibilities have the same chance of occurring; they are **equally likely**.

The probability of each outcome is $\frac{1}{2}$.

$$P(\text{heads}) = P(\text{tails}) = \frac{1}{2}$$

For this spinner, there are 10 possible outcomes in the sample space. The outcomes are equally likely.

$$P(7) = \frac{1}{10} \qquad P(\text{even number}) = \frac{5}{10}, \text{ or } \frac{1}{2}$$

$$P(\text{a number greater than } 4) = \frac{6}{10}, \text{ or } \frac{3}{5}$$

When the possible outcomes are equally likely, you calculate the probability that an event E will occur by using a ratio.

$$P(E) = \frac{\text{number of favorable outcomes}}{\text{total number of possible outcomes}}$$

Find each probability.

1.

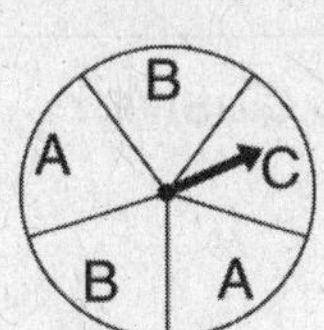

$P(C) = $ ______

$P(A) = $ ______

2.

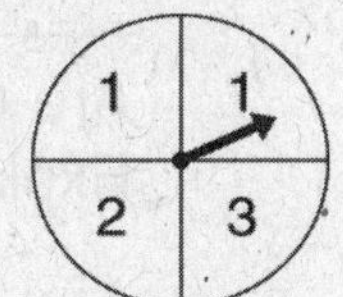

$P(1) = $ ______, or ______

$P(\text{even}) = $ ______

3.

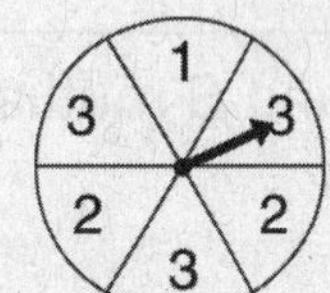

$P(2) = $ ______, or ______

$P(\text{odd}) = $ ______, or ______

Probability can be based upon geometric ratios for lengths, areas, or volumes. This is called **geometric probability.**

The probability that a randomly chosen point inside the triangle is within the square can found using the following.

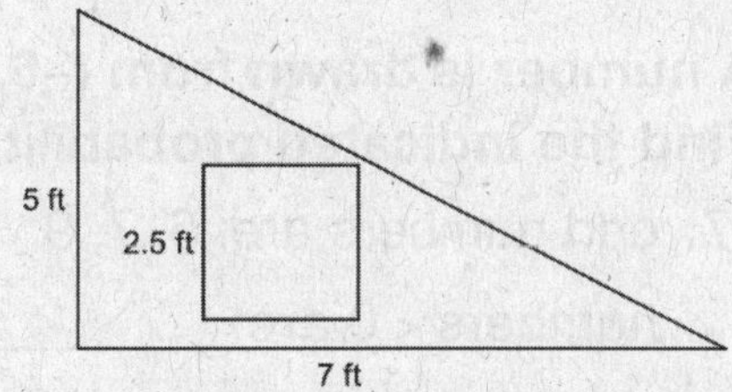

$$\text{probability} = \frac{\text{area of square}}{\text{area of triangle}}$$

4. Using the figure above, find the probability of a point randomly chosen inside the square is within the triangle.

____

Holt McDougal Mathematics

LESSON	**Review for Mastery**
10-3	***Theoretical Probability (continued)***

For this spinner:

$P(\text{odd}) = \dfrac{3}{6}$, or $\dfrac{1}{2}$ $P(\text{even}) = \dfrac{1}{6}$

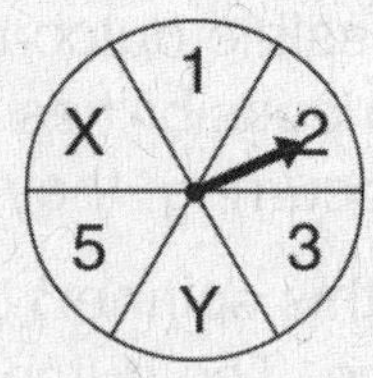

You cannot get an odd number
and an even number $P(\text{odd and even}) = 0$
in the same spin.

Events that cannot occur in the same trial are called **mutually exclusive.**

A number is drawn from {–6, –4, 0, 2, 4, 7, 9}.
List the possible favorable results for each event. Tell if the
events are mutually exclusive.

5. *Event A*: get an odd number

____

Event B: get a negative number

____

Are *A* and *B* mutually exclusive?
Explain.

____

6. *Event C*: get a multiple of 3

____

Event D: get an even number

____

Are *C* and *D* mutually exclusive?
Explain.

____

For the spinner at the top of this page:

$P(\text{odd}) = \dfrac{3}{6}$, or $\dfrac{1}{2}$ $P(\text{even}) = \dfrac{1}{6}$ $P(\text{odd } or \text{ even}) = \dfrac{3}{6} + \dfrac{1}{6} = \dfrac{4}{6}$, or $\dfrac{2}{3}$

A number is drawn from {–6, –4, 0, 5, 6, 7, 9}.
Find the indicated probabilities.

7. odd numbers are: 5, 7, 9

   numbers < 0 are: _______________

   $P(\text{odd}) = \dfrac{}{7}$

   $P(\text{number} < 0) = \dfrac{}{7}$

   $P(\text{odd number or number} < 0) =$

   $\dfrac{}{7} + \dfrac{}{7} = \dfrac{}{7}$

8. numbers > 6 are: _______________

   even numbers: _______________

   $P(\text{number} > 6) =$ _______________

   $P(\text{even number}) =$ _______________

   $P(\text{number} > 6 \text{ or even number}) =$

   _______ + _______ = _______

Holt McDougal Mathematics

LESSON	**Challenge**
10-3	*Picture This*

Venn diagrams can be used to illustrate and solve problem situations involving probability.

Consider a cube numbered 1–6.

Let *Event A* = rolling an even number on the cube.
    favorable outcomes = 2, 4, 6

Let *Event B* = rolling a number less than 5 on the cube.
    favorable outcomes = 1, 2, 3, 4

Note that the numbers 2 and 4 are in both events and, thus, lie in the intersection of the two circles that represent *Events A* and *B*.

So, to determine the probability of getting an even number that is also less than 5, the favorable outcomes are in the intersection of the circles.

$$P(A \text{ and } B) = \frac{\text{number of favorable outcomes}}{\text{total number of possible outcomes}} = \frac{2}{6}, \text{ or } \frac{1}{3}.$$

Then, to determine the probability of getting an even number *or* a number that is less than 5, count the elements found only in circle *A*, only in circle *B*, and in the intersection of circles *A* and *B*.

$$P(A \text{ or } B) = \frac{\text{number of favorable outcomes}}{\text{total number of possible outcomes}} = \frac{5}{6}$$

Draw a Venn diagram to solve each problem.
A cube numbered 1–6 is rolled once.

1. Find the probability of getting an odd number that is greater than 2.

    Event A = a number that is

    Event B = a number that is

    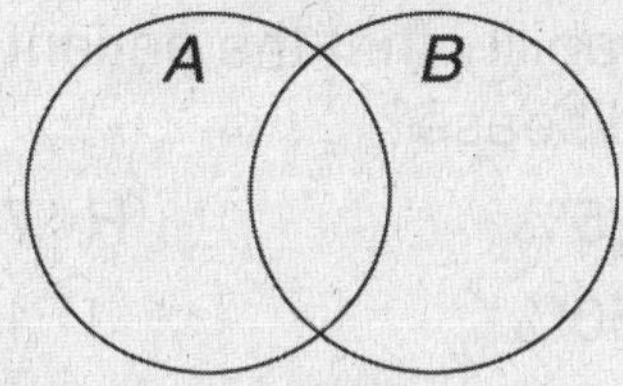

    $P(A \text{ and } B) =$

2. Find the probability of getting an even number that is greater than 3.

    Event A = a number that is

    Event B = a number that is

    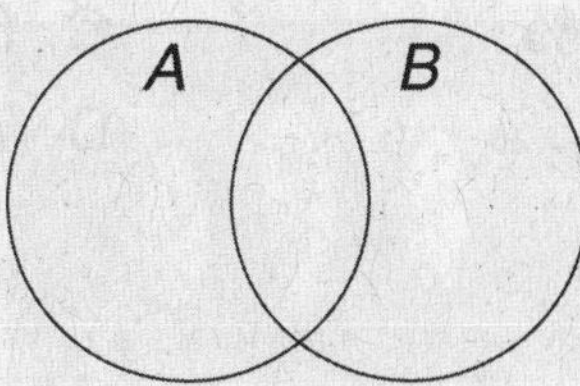

    $P(A \text{ or } B) =$

Holt Mathematics

Problem Solving
LESSON 10-3

Theoretical Probability

A company that sells frozen pizzas is running a promotional special. Out of the next 100,000 boxes of pizza produced, randomly chosen boxes will be prize winners. There will be one grand prize winner who will receive $100,000. Five hundred first prize winners will get $1000, and 3,000 second prize winners will get a free pizza. Write the correct answer in fraction and percent form.

1. What is the probability that the box of pizza you just bought will be a grand prize winner?

   ____

2. What is the probability that the box of pizza you just bought will be a first prize winner?

   ____

3. What is the probability that the box of pizza you just bought will be a second prize winner?

   ____

4. What is the probability that you will win anything with the box of pizza you just bought?

   ____

Researchers recommended that instead of screening all people for certain diseases, they can use a Punnett square to identify the people who are most likely to have the disease. By only screening these people, the cost of screening will be less. Fill in the Punnett square below and use them to choose the letter for the best answer.

5. What is the probability of DD?

   A 0%   C 50%

   B 25%   D 75%

6. What is the probability of Dd?

   F 25%   H 75%

   G 50%   J 100%

7. What is the probability of dd?

   A 0%   C 50%

   B 25%   D 75%

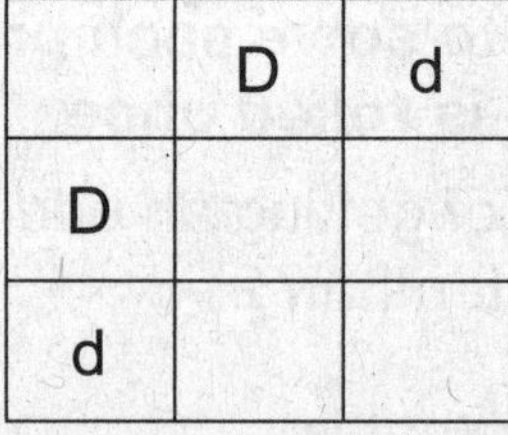

8. DD or Dd indicates that the patient will have the disease. What is the probability that the patient will have the disease?

   F 25%   H 75%

   G 50%   J 100%

Holt McDougal Mathematics

Reading Strategies
Draw Conclusions

Theoretical probability describes what might be expected to happen in an event. It helps you draw conclusions.

$$\text{Probability(event)} = \frac{\text{number of ways an event can occur}}{\text{total number of events}}$$

When you flip a coin, there are two possible events.

→ land on heads

→ land on tails

You have a 1 out of 2 chance for each.

$$P(\text{heads}) = \frac{1}{2} \qquad P(\text{tails}) = \frac{1}{2}$$

These events have the same probability, so you can draw the conclusion that they are **equally likely** to occur.

A number cube has six faces. The numbers on the six faces are 2, 3, 4, 2, 3, and 4. Answer the following about this number cube.

1. How many ways can the number cube land? _______________

2. How many ways can you get a 2 on the cube? _______________

3. What is P(rolling a 2)? _______________

4. How many ways can you get an even number? _______________

5. What is P(rolling an even number)? _______________

6. Can you conclude that P(rolling a 2) and P(rolling an even number) are equally likely to occur? Explain.

____

____

7. What can you conclude about rolling an even number or an odd number?

____

Holt McDougal Mathematics

LESSON 10-3 — Puzzles, Twisters & Teasers
That's Odd!

Using the chart, write the probability for each outcome as a fraction. The order of the coins does not matter. Unscramble the letters to answer the riddle. The fractions under the riddle will give you hints to get you started.

HTT Probability: __________ **E**

THT Probability: __________ **O**

TTT Probability: __________ **P**

HHT Probability: __________ **F**

HHH Probability: __________ **A**

TTH Probability: __________ **S**

HTH Probability: __________ **R**

THH Probability: __________ **K**

TH Probability: __________ **C**

THHT Probability: __________ **Y**

Penny	Dime	Quarter
H	H	H
H	H	T
H	T	T
H	T	H
T	H	H
T	H	T
T	T	H
T	T	T

What did the boy do when he wanted to see time fly?
He dropped his clock

____ ____ **F A**
$\frac{3}{8}$ $\frac{3}{8}$

____ ____ **S** ____ ____ ____ ____ **R**
$\frac{3}{8}$ $\frac{3}{8}$ 0 0 $\frac{3}{8}$ $\frac{1}{8}$ $\frac{1}{8}$ $\frac{3}{8}$

   Holt McDougal Mathematics

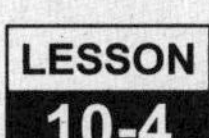

Practice A
LESSON 10-4

Independent and Dependent Events

Determine if the events are dependent or independent.

1. drawing a card from a deck of cards and tossing a coin

   ____

2. drawing two cards from a regular deck of cards and not replacing the first

   ____

3. spinning a number on a spinner and drawing a marble from a container

   ____

4. drawing two red marbles without replacement from a container of red and blue marbles

   ____

An experiment consists of spinning each spinner once. Find the probability. For each spin, all outcomes are equally likely.

5. P(A and 2)

   ____

7. P(B and 3)

   ____

9. P(C and 1 or 2)

   ____

6. P(D and 1)

   ____

8. P(B and 1 or 3)

   ____

10. P(A and not 1)

   ____

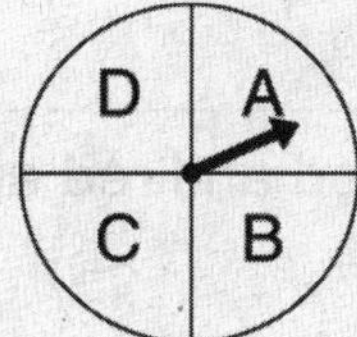

11. Georgiana wants to toss three coins and get all heads. What is the probability of tossing 3 coins and getting 3 heads?

   ____

LESSON 10-4 — Practice B
Independent and Dependent Events

Determine if the events are dependent or independent.

1. choosing a tie and shirt from the closet _________________

2. choosing a month and tossing a coin _________________

3. rolling two fair number cubes once, then rolling them again if you received the same number on both number cubes on the first roll _________________

An experiment consists of rolling a fair number cube and tossing a fair coin.

4. Find the probability of getting a 5 on the number cube and tails on the dime. _________

5. Find the probability of getting an even number on the number cube and heads on the dime. _________

6. Find the probability of getting a 2 or 3 on the number cube and heads on the dime. _________

A box contains 3 red marbles, 6 blue marbles, and 1 white marble. The marbles are selected at random, one at a time, and are not replaced. Find the probability.

7. P(blue and red)

8. P(white and blue)

9. P(red and white)

10. P(red and white and blue)

11. P(red and red and blue)

12. P(red and blue and blue)

13. P(red and red and red)

14. P(white and blue and blue)

15. P(white and red and white)

Holt McDougal Mathematics

<table><tr><td>LESSON
10-4</td><td></td></tr></table>

Practice C

Independent and Dependent Events

Consider a regular deck of cards without the jokers. Cards are replaced after each draw. Find the probability of each of the following.

1. P(pair of red kings)

2. P(a diamond and a black seven)

Use the same deck of cards but do not replace the card after each draw.

3. P(ace of hearts and king of hearts)

4. P(a ten and a jack)

5. P(red card and a black card)

6. P(a club and king or red ace)

7. Mr. and Mrs. Reginald are expecting their first baby. The doctor tells them they are having triplets. What is the probability that the babies will all be the same sex?

8. Sid has a bag of 12 red, 14 brown, and 10 blue marbles. He chooses one, shoots it, and chooses another. What is the probability that his first selection is a red marble, and then a blue marble?

9. If Justine's initials are JMD, what is the probability that she will draw her initials from a box containing the letters of the alphabet? There is no replacement of letters after each is drawn.

10. There are 13 math students, 10 science students, and 17 English students in a group. If only one prize is allowed per person, what is the probability that the moderator will award a science student a prize and then award another prize to a math student?

Holt McDougal Mathematics

LESSON 10-4

Review for Mastery

Independent and Dependent Events

Carlos is to draw 2 straws at random from a box of straws that
contains 4 red, 4 white, and 4 striped straws.

P(1st straw is striped) = $\dfrac{4}{12}$ ⟵ number of striped straws
⟵ total number of straws

If Carlos *returns* the 1st straw to the box before drawing the 2nd straw, the probability that the 2nd straw is striped remains the same.	If Carlos *does not return* the first straw to the box before drawing the second straw, the probability that the second straw is striped changes.

P(2nd straw is striped)
= $\dfrac{4}{12}$ ⟵ same number of striped straws
⟵ same total number of straws

When the 1st straw is returned before
the 2nd draw, the 2nd draw occurs as
though the 1st draw never happened,
independent events.

P(striped and striped) = $\dfrac{4}{12} \times \dfrac{4}{12}$

= $\dfrac{1}{3} \times \dfrac{1}{3} = \dfrac{1}{9}$

P(2nd straw is striped)
= $\dfrac{3}{11}$ ⟵ one striped straw has been taken
⟵ one less straw in total number

When the 1st straw is not returned before
the 2nd draw, the number of straws
remaining is changed, **dependent
events**.

P(striped and striped) = $\dfrac{4}{12} \times \dfrac{3}{11}$

= $\dfrac{1}{3} \times \dfrac{3}{11} = \dfrac{1}{11}$

Describe the events as independent or dependent.

1. Josh tosses a coin and spins a spinner. _______________

2. Ana draws a colored toothpick from a jar.
   Without replacing it, she draws a second toothpick. _______________

3. Sue draws a card from a deck of cards and replaces it.
   Then she draws a second card from the deck. _______________

**Each situation begins with a box of marbles that contains 2 red, 3 blue,
4 green, and 3 yellow marbles. Complete to find each probability.**

4. A 1st marble is drawn and replaced.
   Then a 2nd marble is drawn.

   P(red and blue) = $\dfrac{}{12} \times \dfrac{}{12} =$ _______

5. A 1st marble is drawn and not
   replaced. A 2nd marble is drawn.

   P(red and blue) = $\dfrac{}{12} \times \dfrac{}{11} =$ _______

6. A 1st marble is drawn and replaced.
   Then a 2nd marble is drawn.

   P(red and red) = $\dfrac{}{12} \times \dfrac{}{} =$ _______

7. A 1st marble is drawn and not
   replaced. A 2nd marble is drawn.

   P(red and red) = $\dfrac{}{12} \times \dfrac{}{} =$ _______

   Holt McDougal Mathematics

LESSON
10-4

Challenge

Probability from a Table

The table shows the results of a survey of 50 students. The students were asked whether they liked a newly released movie.

	Yes	No
Male	16	14
Female	12	8

According to the table:

    16 + 14, or 30 males were surveyed.

    12 + 8, or 20 females were surveyed.

    16 + 12, or 28, of those surveyed liked the movie.

    14 + 8, or 22, of those surveyed did not like the movie.

The probability that a student randomly selected from the group is a male is $\frac{30}{50}$, or $\frac{3}{5}$.

The probability that a student did not like the movie is $\frac{22}{50}$, or $\frac{11}{25}$.

The probability that a student is a male who did not like the movie is $\frac{14}{50}$, or $\frac{7}{25}$.

The table below shows the results of a survey of 50 students. The students were asked whether they liked a newly released CD. One student was selected at random from this group. Use the table to solve.

	Yes	No
Male	35	5
Female	8	2

1. What is the probability that the selected student liked the CD? _______________

2. What is the probability that the student did not like the CD? _______________

3. What is the probability that the student is a female? _______________

4. What is the probability that the student is a female who liked the CD? _______________

5. What is the probability that the student is a male who did not like the CD? _______________

Holt McDougal Mathematics

Problem Solving

Independent and Dependent Events

Are the events independent or dependent? Write the correct answer.

1. Selecting a piece of fruit, then choosing a drink.

   ____

2. Buying a CD, then going to another store to buy a video tape if you have enough money left.

   ____

Dr. Fred Hoppe of McMaster University claims that the probability of winning a pick 6 number game where six numbers are drawn from the set 1 through 49 is about the same as getting 24 heads in a row when you flip a fair coin.

3. Find the probability of winning the pick 6 game and the probability of getting 24 heads in a row when you flip a fair coin.

   ____

4. Which is more likely: to win a pick 6 game or to get 24 heads in a row when you flip a fair coin?

   ____

In a shipment of 20 computers, 3 are defective. Choose the letter for the best answer.

5. Three computers are randomly selected and tested. What is the probability that all three are defective if the first and second ones are not replaced after being tested?

   A $\dfrac{1}{760}$ C $\dfrac{27}{8000}$

   B $\dfrac{1}{1140}$ D $\dfrac{3}{5000}$

6. Three computers are randomly selected and tested. What is the probability that all three are defective if the first and second ones are replaced after being tested?

   F $\dfrac{1}{760}$ H $\dfrac{27}{8000}$

   G $\dfrac{1}{1140}$ J $\dfrac{3}{5000}$

7. Three computers are randomly selected and tested. What is the probability that none are defective if the first and second ones are not replaced after being tested?

   A $\dfrac{34}{57}$ C $\dfrac{4913}{6840}$

   B $\dfrac{4913}{8000}$ D $\dfrac{1}{2000}$

8. Three computers are randomly selected and tested. What is the probability that none are defective if the first and second ones are replaced after being tested?

   F $\dfrac{34}{57}$ H $\dfrac{4913}{6840}$

   G $\dfrac{4913}{8000}$ J $\dfrac{1}{2000}$

Holt McDougal Mathematics

Reading Strategies

LESSON 10-4

Use Context

When the outcome of one event does not affect the outcome of another, they are called **independent events.**

A cube has the numbers 1–6 on the faces.

P(rolling a 6) is $\dfrac{1}{6}$.

If you roll the cube a second time, P(rolling a 6) is still $\dfrac{1}{6}$.

The second roll of the cube is independent of the first roll.

When the outcome of one event affects the probability of another, they are called **dependent events.**

There are 10 cubes in a bag. Three of them are red.

P(drawing a red cube) is $\dfrac{3}{10}$.

If a red cube is drawn and not replaced, the probability of drawing a red cube on the second draw is 2 out of 9.

So P(drawing another red cube) is $\dfrac{2}{9}$.

Drawing a cube the second time is dependent on drawing the first cube.

Tell if each situation describes dependent events or independent events.

1. You flip a coin. Then you flip a coin a second time.

   ____

2. You roll a number cube three times.

   ____

3. You draw a card from a deck of cards and do not replace the card. Then you draw a second time.

   ____

4. You spin a spinner once. Then you spin the spinner a second time.

   ____

5. You pull a marble out of a jar and leave it on the table. Then you pull another marble out of the jar.

   ____

Holt McDougal Mathematics

Puzzles, Twisters & Teasers

LESSON 10-4

Declare Your Independence!

**Decide whether the events are dependent or independent.
Circle the letter above your answer. Unscramble the letters to
answer the riddle.**

1. tossing a coin twice

   Q   **I**

   dependent   independent

2. drawing names out of a hat

   R   **B**

   dependent   independent

3. pulling two socks from a drawer
   at the same time

   E   **P**

   dependent   independent

4. throwing a pair of number cubes ten
   times

   X   **U**

   dependent   independent

5. drawing two marbles out of a bag

   B   **D**

   dependent   independent

6. throwing one number cube five times

   U   **S**

   dependent   independent

7. spinning a spinner five times

   M   **C**

   dependent   independent

8. picking cards from a deck

   G   **V**

   dependent   independent

9. spinning two different spinners two
   times

   K   **E**

   dependent   independent

10. throwing three coins three times

   A   **R**

   dependent   independent

What do penguins eat for lunch?

_____ _____ _____ _____ _____   _____ _____ _____ _____ _____ _____

Family Letter

10B Probability and Counting

Dear Family,

The student will learn how to apply the concepts of probability.

The student will be introduced to the **Fundamental Counting Principle** to find the total number of ways that two or more separate events can happen.

A game consists of rolling a number cube and spinning a spinner divided into equal quarters of different colors. How many different outcomes in one turn are possible?

Number of outcomes when rolling a number cube: 6
Number of outcomes when spinning the spinner: 4
6 • 4 = 24. There are 24 possible outcomes in any one turn.

The student will also learn how to use **tree diagrams.**

At a restaurant, Jane has a choice of 2 appetizers, 3 entrées, and 2 desserts. How many ways can she choose an appetizer, an entrée, and a desert?

Appetizer	Entrée	Dessert
Soup	Chicken	Fruit Salad
Salad	Beef	Parfait
	Veggie	

Make a tree diagram.

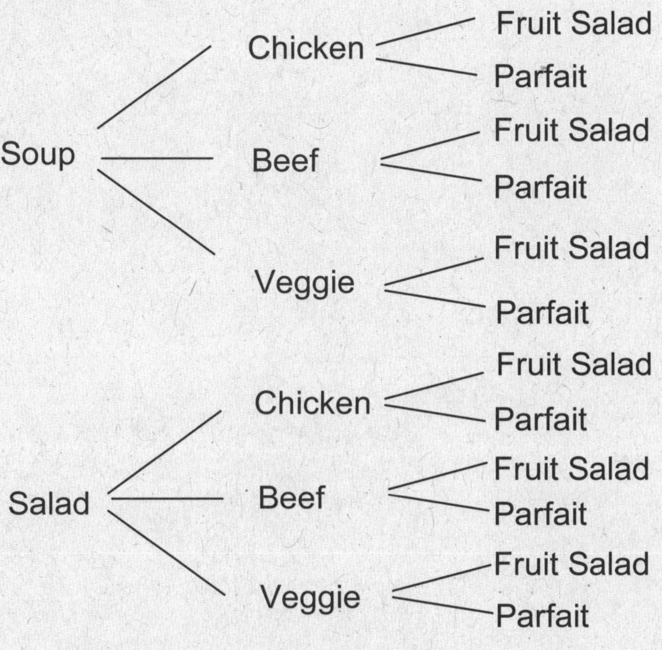

Jane can choose an appetizer, an entrée, and a dessert 12 different ways.

Vocabulary

These are the math words we are learning:

combination an arrangement of items or events in which order does not matter

factorial the product of all the whole numbers from a given number down to 1

Fundamental Counting Principle if there are *m* ways to choose a first item and *n* ways to choose a second item after the first item has been chosen, then there are *m • n* ways to choose both items

permutation an arrangement of events in a certain order

tree diagram a branching diagram that shows all the possible outcomes of an event

Holt McDougal Mathematics

Family Letter

10B Probability and Counting *continued*

The student will learn the difference between **permutations** and **combinations** and how to evaluate each.

You are required to read 3 books this summer. You can choose from a list of 10 books. How many different ways can you choose the 3 books?

Since order does not matter, this is a combination.

$$_{10}C_3 = \frac{10!}{3!(10-3)!} = \frac{10 \cdot 9 \cdot 8 \cdot 7 \cdot 6 \cdot 5 \cdot 4 \cdot 3 \cdot 2 \cdot 1}{3 \cdot 2 \cdot 1 \cdot 7 \cdot 6 \cdot 5 \cdot 4 \cdot 3 \cdot 2 \cdot 1} = 120$$

If you must read one in June, one in July, and one in August, how many different ways can you do this?

Since order does matter, this is a permutation.

$$_{10}P_3 = \frac{10!}{(10-3)!} = \frac{10 \cdot 9 \cdot 8 \cdot 7 \cdot 6 \cdot 5 \cdot 4 \cdot 3 \cdot 2 \cdot 1}{7 \cdot 6 \cdot 5 \cdot 4 \cdot 3 \cdot 2 \cdot 1} = 720$$

The student may have difficulty deciding whether order is important when calculating combinations and permutations. Try to find examples in real-world situations and discuss with the student whether it is a combination or permutation. Using the skills learned here in everyday situations will reinforce them with the student.

Sincerely,

 Holt McDougal Mathematics

At-Home Practice

10B Probability and Counting

Find the number of possible outcomes.

1. A pizza shop offers thick crusts, thin crusts, and stuffed crusts. The choices of toppings are pepperoni, cheese, hamburger, peppers, sausage, onions and mushrooms. How many different one-topping pizzas can you order? _______________

2. If a code consists of 3 letters, followed by 2 numbers, how many different codes can be made? _______________

Evaluate each expression.

3. $_6C_2$

____

4. $_8P_3$

____

Tell whether each problem requires a combination or a permutation, then solve.

5. Mr. Jaspers is choosing 3 students to solve a problem on the chalkboard. How many ways can he choose 3 students out of the 24 in his class?

____

6. Dr. Fleming has 4 open appointment times at 9:00, 9:30. 10:00, and 10:30. If she has 7 patients who need appointments, how many ways can she fill the appointments?

____

7. There are 10 teams in a bowling league, and 4 will make it to the playoffs. How many 4-team groups can make the playoffs?

____

Answers: 1. $3 \cdot 7 = 21$ **2.** $26 \cdot 26 \cdot 26 \cdot 10 \cdot 10 = 1{,}757{,}600$ **3.** 15 **4.** 336 **5.** combination; 2024 **6.** permutation; 840 **7.** combination; 210

<table>
<tr><td>CHAPTER 10</td><td><h1>Family Fun</h1>Quick Combinations</td></tr>
</table>

Directions

- Cut out the cards below and place them face up on a table.

- One player at a time will try and make 10 possible 3-number combinations out of the numbers below. Another player should record the combinations in the table to the right as well as the time.

- The player should keep making combinations until they find all 10. Remember, order does not matter in combinations, so 123 is the same as 321, etc.

- The player with the fastest time wins the game.

	Player 1	Player 2	Player 3
1			
2			
3			
4			
5			
6			
7			
8			
9			
10			
Time			

$$1 \quad 2$$

$$3 \quad 4 \quad 5$$

Answer: The ten different combinations in any order are: 123, 124, 125, 134, 135, 145, 234, 235, 245, 345

Holt McDougal Mathematics

<table><tr><td>**LESSON**
10-5</td><td># Practice A
Making Decisions and Predictions</td></tr></table>

The zoo store sells caps with different animals pictured on the cap. The table shows the animals pictured on the last 100 caps sold. The manager plans to order 1500 new caps.

Animal Caps Sold

Animal	Number
Tiger	30
Orangutan	20
Panda Bear	25
Giraffe	18
Gazelle	7

1. Find the probability of selling a tiger cap.

   ____

2. How many tiger caps should the manager order?

   ____

3. Find the probability of selling a panda bear cap.

   ____

4. How many panda bear caps should the manager order?

   ____

5. Use probability to decide how many orangutan caps the manager should order.

   ____

6. Nancy spins the spinner at the right 60 times. Predict how many times the spinner will land on the number 2.

   ____

Decide whether the game is fair.

7. Roll two fair number cubes labeled 1–6. Player A wins if both numbers are odd.
   Player B wins if both numbers are even.

   ____

   Holt McDougal Mathematics

LESSON 10-5 Practice B

Making Decisions and Predictions

A sports store sells water bottles in different colors. The table shows the colors of the last 200 water bottles sold. The manager plans to order 1800 new water bottles.

Water Bottles Sold

Color	Number
Red	30
Blue	50
Green	25
Yellow	10
Purple	10
Clear	75

1. How many red water bottles should the manager order? _______________

2. How many green water bottles should the manager order? _______________

3. How many clear water bottles should the manager order? _______________

4. If the carnival spinner lands on 10, the player gets a large stuffed animal. Suppose the spinner is spun 30 times. Predict how many large stuffed animals will be given away. _______________

Decide whether the game is fair.

5. Roll two fair number cubes labeled 1–6. Player A wins if both numbers are the same. Player B wins if both numbers are different.

____

6. Roll two fair number cubes labeled 1–6. Add the numbers. Player A wins if the sum is 5 or less. Player B wins if the sum is 9 or more.

____

7. Toss three fair coins. Player A wins if exactly one tail lands up. Otherwise, Player B wins.

____

  Holt McDougal Mathematics

LESSON 10-5

Practice C

Making Decisions and Predictions

A fair number cube is labeled 1–6. Predict the number of outcomes for the given number of rolls.

1. outcome: 5
   number of rolls: 42

   ____

2. outcome: less than 3
   number of rolls: 75

   ____

3. outcome: not 4
   number of rolls: 60

   ____

4. outcome: 2, 3, or 4
   number of rolls: 30

   ____

5. outcome: greater than 2
   number of rolls: 48

   ____

6. outcome: multiple of 3
   number of rolls: 90

   ____

7. In his last eight 5K runs, Jeremy had the following times in minutes: 24:48, 23:45, 23:12, 24:08, 25:36, 22:03, 23:29, and 24:01. Based on these results, what is the best prediction of the number of times Jeremy will run faster than 24 minutes in his next 20 5K runs?

   ____

8. Before a school vote on a mascot for a community river project, a sample of students surveyed gave the otter 18 votes, the osprey 8 votes, and the beaver 6 votes. Based on these results, predict the number of votes for each animal if 1200 students vote.

   ____

Decide whether each game is fair.

9. A spinner is divided evenly into 6 sections. There are 3 green sections, 2 blue sections, and 1 yellow section. Player A wins if the spinner does not land on green. Otherwise, Player B wins.

   ____

10. Roll two fair number cubes labeled 1–6. Add the numbers. Player A wins if the sum is 8 or more. Player B wins if the sum is 5 or less.

    ____

11. Toss three fair coins. Player A wins if exactly two tails land up. Player B wins if all heads or all tails land up.

    ____

   Holt McDougal Mathematics

| LESSON 10-5 | **Review for Mastery**
Making Decisions and Predictions |

Probability can be used to make predictions about data.

The spinner has 5 equal sections. To predict how many times you will land on the number 1 in 30 spins, first find the probability of landing on 1 in one spin. Then multiply 30 spins by that probability.

Step 1 Find the probability.

$$P(1) = \frac{\text{number of 1s}}{\text{number of equal sections}} = \frac{2}{5}$$

Step 2 Multiply the number of spins by the probability.

There are 30 spins and $P(1) = \frac{2}{5}$.

$$30 \times \frac{2}{5} = 6 \times 2 = 12$$

You will spin about 12 1s in 30 spins.

Each situation begins with a box of marbles that contains 2 red, 3 blue, 4 green, and 3 yellow marbles. Complete to find each probability.

1. Predict how many times you will land on C in 32 spins.

   a. $P(C) =$ ___________

   b. $32 \times P(C) = 32 \times$ ___________ $=$ ___________

   c. about ___________ times

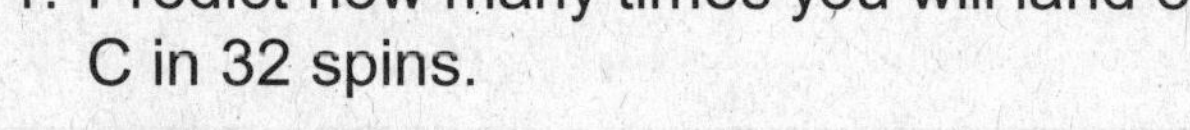

2. Predict how many times you will land on B in 48 spins.

   a. $P(B) =$ ___________

   b. $48 \times P(B) =$ ___________ $=$ ___________

   c. about ___________ times

3. Predict how many times you will land on A in 50 spins.

   a. $P(A) =$ ___________

   b. ___________ $=$ ___________

   c. about ___________ times

Holt McDougal Mathematics

<table>
<tr><td>LESSON
10-5</td><td></td></tr>
</table>

Challenge

Complimentary Events

The complement of an event A is the event that A does not occur. It includes all outcomes not in A. Represent the complement of A as not A.

Since either A or not A must occur, $P(A) + P(\text{not } A) = 1$.
This means that $P(A) = 1 - P(\text{not } A)$ and $P(\text{not } A) = 1 - P(A)$.

Suppose A is the event rolling 4 when rolling a fair number cube. Then the complement of A is the event rolling a number that is not 4.

$$P(4) = \frac{1}{6} \qquad P(\text{not } 4) = 1 - \frac{1}{6} = \frac{5}{6}$$

Two fair number cubes are rolled. Solve.

1. Suppose A is the event rolling a sum of 2 or 3.

   a. Find $P(A) =$ _______________

   b. Use words to describe the complement of A.

   ____

   c. Find $P(\text{not } A)$ _______________

2. Suppose A is the event rolling a sum of at least 10.

   a. $P(A) =$ _______________

   b. Use words to describe the complement of A.

   ____

   c. Find $P(\text{not } A)$. _______________

3. Suppose A is the event rolling a sum of 8.

   a. $P(A) =$ _______________

   b. Use words to describe the complement of A.

   ____

   c. Find $P(\text{not } A)$. _______________

Holt McDougal Mathematics

<table><tr><td>LESSON
10-5</td><td></td></tr></table>

Problem Solving
Making Decisions and Predictions

Write the correct answer.

1. A quality control inspector at a light bulb factory finds 2 defective bulbs in a batch of 1000 light bulbs. If the plant manufactures 75,000 light bulbs in one day, predict how many will be defective.

2. A game consists of rolling two fair number cubes labeled 1–6. Add both numbers. Player A wins if the sum is greater than 10. Player B wins if the sum is 7. Is the game fair or not? Explain.

3. A spinner has 5 equal sections numbered 1–5. Predict how many times Kevin will spin an even number in 40 spins.

4. In her last six 100-meter runs, Lee had the following times in seconds: 12:04, 13:11, 12:25, 11:58, 12:37, and 13:20. Based on these results, what is the best prediction of the number of times Lee will run faster than 13 seconds in her next 30 runs?

Use the table below that shows the number of colors of the last 200 T-shirts sold at a T-shirt shop. The manager of the store wants to order 1800 new T-shirts. Choose the letter of the best answer

5. How many red T-shirts should the manager order?

   A 175   C 378

   B 315   D 630

6. How many blue T-shirts should the manager order?

   F 495   H 900

   G 665   J 990

7. How many more black T-shirts than white T-shirts should the manager order?

   A 855   C 315

   B 585   D 270

T-Shirts Sold

Color	Number
Red	35
Blue	55
Green	15
Black	65
White	30

Holt McDougal Mathematics

Reading Strategies
Draw Conclusions

You can use probability to make decisions about data.

The table shows the colors of 100 wrist bands sold at a sports shop.

Wrist Bands Sold

Color	Number
Yellow	40
Blue	20
Pink	30
Red	10

The store manager wants to order 2000 new bands. She can use the data in the table to decide how many of each color of wrist band to order in the next order. The manager should order the greatest number of yellow bands and the fewest number of red bands.

To find the number of red bands, first use the data in the table to find $P(red)$, the probability of selling a red band.

$$P(red) = \frac{10}{100} = 0.1$$

Then use the probability to find how many red bands to order:
0.1 of 2000 = 0.1 × 2000 = 200

So, the manager should order 200 red bands.

Answer each question. Use the information above.

1. What is the probability of selling a yellow band?

   ____

2. Write an expression that represents how many yellow bands the manager should order.

   ____

3. How many yellow bands should the manager order?

   ____

4. What is the probability of selling a pink band?

   ____

5. How many pink bands should the manager order?

   ____

 Holt McDougal Mathematics

Puzzles, Twisters & Teasers

LESSON 10-5

A Friendly Solution!

Find and circle words from the list in the word search horizontally, vertically, or diagonally. Find a word that answers the riddle. Circle it and write it on the line.

| probability | decision | prediction | fair | unfair |
| game | proportion | chance | likely | favorable |

```
H G R E C O M D A R E I G N P
A A D I T C V E I G V E R S R
F V G U E H T C S T A B E S E
A E L N V A T I L O S M A E D
V F R I E N D S H I P A E L I
O S A R L C O I P O K D E R C
R N O I P E S O T E L E R V T
A P C H R S U N F A I R L T I
B R P R O P O R T I O N A Y O
L O V P R O B A B I L I T Y N
E C H A F A V N C E B I L U N
```

What kind of ship never sinks? _______________________________

  Holt McDougal Mathematics

Practice A

LESSON 10-6

The Fundamental Counting Principle

1. A snack bar serves tea, juice, and milk in small, medium, and large sizes. List all the different possible beverage orders.

   ____

   ____

2. The school's football team has a choice of different colored jerseys and different colored pants to wear for their uniforms. They have purple jerseys, white jerseys, and striped jerseys. The choices for the pants are purple or white. List all the different possible uniforms the team can wear.

   ____

   ____

   ____

   ____

   ____

3. What is the probability that the team will randomly select the white jerseys with purple pants?

   ____

4. Student identification codes at a high school are 4-digit randomly generated codes beginning with 1 letter and ending with 3 numbers. If all codes are equally likely, how many possible codes are there?

   ____

5. Find the probability of being assigned the code A123.

   ____

6. There are 6 shoes in a closet, 4 blue and 2 red. Two shoes are randomly selected. What is the probability that a matching pair is chosen?

   ____

Holt McDougal Mathematics

<table><tr><td>LESSON
10-6</td><td>

Practice B
The Fundamental Counting Principle
</td></tr></table>

Employee identification codes at a company contain 2 letters followed by 2 numbers. All codes are equally likely.

1. Find the number of possible identification codes.

2. Find the probability of being assigned the code MT49.

3. Find the probability that an ID code of the company does not contain the letter A as the second letter of the code.

4. Find the probability that an ID code of the company does not contain the number 2.

5. Mrs. Sharpe is planning her dinners for next week. The choices for the entree are roast beef, turkey, or pork. The choices of carbohydrates are mashed potatoes, baked potatoes, or noodles. The vegetable choices are broccoli, spinach, or carrots. Make a tree diagram indicating the possible outcomes for each entree.

6. How many different meals could Mrs. Sharpe prepare? _______________

Find the probability for each of the following.

7. *P*(dinner with baked potato)

8. *P*(dinner with noodles and carrots)

9. There are 10 gloves in a drawer, 6 leather and 4 cotton. Two gloves are drawn. What is the probability the gloves match?

Holt McDougal Mathematics

LESSON 10-6	# Practice C

The Fundamental Counting Principle

Find the number of possible outcomes.

1. pasta: spaghetti, linguine

   sauce: pesto, Alfredo, marinara

   ____

2. music: country, pop, rap

   artist: male, female, duo, group

   ____

3. eye color: blue, brown, green

   hair color: black, blond, brown, red

   sex: male, female

   ____

4. font: Arial, Calligraphy, Helvetica

   size: 10, 12, 14, 16, 20, 22, 24

   color: black, red, blue, green

   ____

5. sport: baseball, basketball, football, hockey, soccer, volleyball

   level: professional, college, high school, grade school

   ____

Use the chart for Exercises 6 and 7.

Main Color	Trim Color	Frame Size	Tire Size	Gears
blue	white	19 in.	24 in.	15 speed
green	black	21 in.	26 in.	24 speed
red	yellow	23 in.		

6. Janis plans to buy a bike. How many combinations are
   possible with a choice of one main color, one trim color,
   one frame size, one tire size, and one gear selection?   _______________________

7. Janis decides to buy a green bike. How many combinations
   are now possible?   _______________________

**A computer randomly generates a 5-character password of
3 letters followed by 2 digits. All passwords are equally likely.**

8. Find the probability that a password contains exactly one 2.   _______________________

9. Find the probability that a password contains exactly one *A*.   _______________________

          Holt McDougal Mathematics

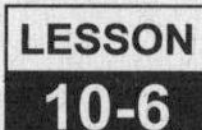

LESSON 10-6 · Review for Mastery
The Fundamental Counting Principle

The Fundamental Counting Principle can help you solve some problems about situations that involve more than one activity.

| **the number of ways in which one activity can be performed** | × | **the number of ways in which a second activity can be performed** | = | **the total number of ways in which both activities can be performed** |

Apply the Fundamental Counting Principle to find the total number of possibilities in each situation.

1. Kelly has 6 shirts and 4 coordinating pants. The number of possible shirt-pants outfits is: _____________, or _________

2. The menu for dinner lists 2 soups, 4 meats, and 3 desserts. How many different meals that have one soup, one meat, and one dessert are possible? _____________, or _________

A **tree diagram** helps you see all the possibilities in a sample space.

If three coins are tossed at the same time, list all the possible outcomes.

List, in a column, the 2 possibilities for the 1st coin.

For each possibility for the 1st coin, list the 2 possibilities for the 2nd coin.

For each possibility for the 2nd coin, list the 2 possibilities for the 3rd coin.

Read the diagram across to write the list of all possible outcomes.

In this situation, there are 2 × 2 × 2 = 8 possible outcomes.

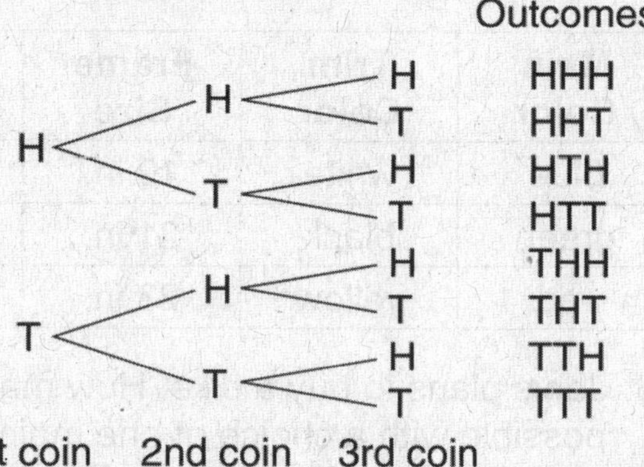

Draw a tree diagram and list the outcomes.

3. A vendor is selling cups of ice cream. There are 2 different sizes of cups: small (S), or large (L). There are 2 different flavors of ice cream: vanilla (V) or chocolate (C). There are 2 different toppings: fudge (F) or pineapple (P).

Holt McDougal Mathematics

<table>
<tr><td>LESSON
10-6</td><td>

Review for Mastery
The Fundamental Counting Principle (continued)

</td></tr>
</table>

How many different 5-letter "words" are possible using the letters of TRIANGLE? Letters can be used only once in each "word."

There are 8 choices for the 1st letter, 7 choices for the 2nd letter,

6 choices for the 3rd letter, 5 choices for the 4th letter, and 4 choices for the 5th.

Apply the Fundamental Counting Principle.

$$\underset{\text{1st letter}}{8} \times \underset{\text{2nd letter}}{7} \times \underset{\text{3rd letter}}{6} \times \underset{\text{4th letter}}{5} = \underset{\text{5th letter}}{4} = 6720 \text{ possibilities}$$

If a "word" is selected at random from the 6720 possibilities, what is the probability that it will be the "word" ANGLE?

There is only one outcome ANGLE. $P(\text{ANGLE}) = \dfrac{1}{6720}$

If a "word" is selected at random from the 6720 possibilities, what is the probability that it will not contain the letter G?

Find the number of favorable outcomes.

Eliminate the letter G from the choices. So, there are 7 choices to begin.

$$\underset{\text{1st letter}}{7} \times \underset{\text{2nd letter}}{6} \times \underset{\text{3rd letter}}{5} \times \underset{\text{4th letter}}{4} \times \underset{\text{5th letter}}{3} = 2520 \text{ possibilities}$$

$$P(\text{5-letter "word" with no G}) = \frac{\text{number of favorable outcomes}}{\text{total number of possible outcomes}} = \frac{2520}{6720}, \text{ or } \frac{3}{8}$$

Apply the Fundamental Counting Principle.

4. Consider the letters of the word MEDIAN.

   a. How many different 4-letter "words" are possible?
      Letters can be used only once.

   $$\underset{\text{1st letter}}{\underline{\quad\quad}} \times \underset{\text{2nd letter}}{\underline{\quad\quad}} \times \underset{\text{3rd letter}}{\underline{\quad\quad}} \times \underset{\text{4th letter}}{\underline{\quad\quad}} = \underline{\quad\quad} \text{ possibilities}$$

   b. If a 4-letter "word" is selected at random
      from all the possibilities, then: $P(\text{DEAN}) = \underline{\quad\quad}$

   c. If a 4-letter "word" is selected at random from all the possibilities,
      what is the probability that it will not contain the letter D?

   $$\text{favorable outcomes: } \underset{\text{1st letter}}{\underline{\quad\quad}} \times \underset{\text{2nd letter}}{\underline{\quad\quad}} \times \underset{\text{3rd letter}}{\underline{\quad\quad}} \times \underset{\text{4th letter}}{\underline{\quad\quad}} = \underline{\quad\quad}$$

   $$P(\text{4-letter "word" with no D}) = \frac{\text{number of favorable outcomes}}{\text{total number of possible outcomes}} = \underline{\quad\quad}, \text{ or } \underline{\quad\quad}$$

                                                   Holt McDougal Mathematics

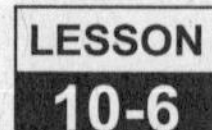

LESSON 10-6

Challenge

Answer the Phone

The world is divided into 9 telephone numbering zones. The North American Numbering Plan (NANP) was developed in 1947 to enable direct dialing without the need for an operator.

NANP numbers are 10 digits in length, of the form

$$N\,X\,X - N\,X\,X - X\,X\,X\,X$$

area code   prefix   line number

Originally, the plan created 86 areas and allowed for expansion to 144 areas. In 1995, NANP expanded to 792 area codes.

1. For the 3-digit area code NXX, the plan allows N to be any digit 2–9. Currently, there are no restrictions on the other 2 digits of the area code. How many area codes are possible?

   ____

2. For the 3-digit prefix, the plan allows N to be any digit 2–9. How many line numbers are possible for a given prefix?

   ____

3. How many telephone numbers are possible for a given area code?

   ____

Some of the prefixes are reserved for services. They are of the form N11 where N is any digit 2–9.

The most familiar service code is 911, reserved for emergency calls. Other currently assigned service codes are 411 (local directory assistance). 611 (repairs), 711 (teletypewriter [hearing/speech impaired]), 811(business office).

4. If all the service code prefixes are removed, how many telephone numbers are possible for a given area code?

   ____

Some other prefixes are not available for general use, such as:

555 (information), 800 and 888 (usually, but not always, toll free), 900 (pay per call).

5. For each prefix that is not available for general use, how many fewer telephone numbers are available for general use?

   ____

Holt McDougal Mathematics

Problem Solving

The Fundamental Counting Principle

Write the correct answer.

1. The 5-digit zip code system for United States mail was implemented in 1963. How many different possibilities of zip codes are there with a 5-digit zip code where each digit can be 0 through 9?

2. In 1983, the ZIP +4 zip code system was introduced so mail could be more easily sorted by the 5-digit zip code plus an additional 4 digits. How many different possibilities of zip codes are there with the ZIP +4 system?

3. In Canada, each postal code has 6 symbols. The first, third and fifth symbols are letters of the alphabet and the second, fourth and sixth symbols are digits from 0 through 9. How many possible postal codes are there in Canada?

4. In the United Kingdom the postal code has 6 symbols. The first, second, fifth and sixth are letters of the alphabet and the third and fourth are digits from 0 through 9. How many possible postal codes are there in the United Kingdom?

Choose the letter for the best answer.

5. In a city in Kansas, all of the phone numbers begin 852–4. The only differences in the phone numbers are the last 3 digits. How many possible phone numbers can be assigned using this system?

   A 729
   B 1000
   C 6561
   D 10,000

6. Many large cities have run out of phone numbers and so a new area code must be introduced. How many different phone numbers are there in a single area code if the first digit can't be zero?

   F 90,000
   G 4,782,969
   H 9,000,000
   J 10,000,000

7. How many different phone numbers are possible using a 3-digit area code and a 7-digit phone number if the first digit of the area code and phone number cannot be zero?

   A 3,486,784,401
   B 8,100,000,000
   C 9,500,000,000
   D 10,000,000,000

8. A shipping service offers to send packages by ground delivery using 2 different companies, by next day air using 3 different companies, and by 2-day air using 3 different companies. How many different shipping options does the service offer?

   F 3
   G 8
   H 10
   J 18

Holt McDougal Mathematics

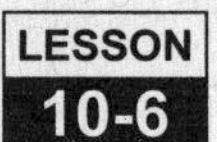

Reading Strategies
Use a Visual Aid

If you have 2 pairs of shorts and 3 shirts, how many different outfits can you make?

Shorts A   Shorts B   Shirt A   Shirt B   Shirt C

A **tree diagram** is a way to visualize the possible outfits.

2 pairs of shorts → Each paired with one → Number of
of the 3 shirts   different outfits

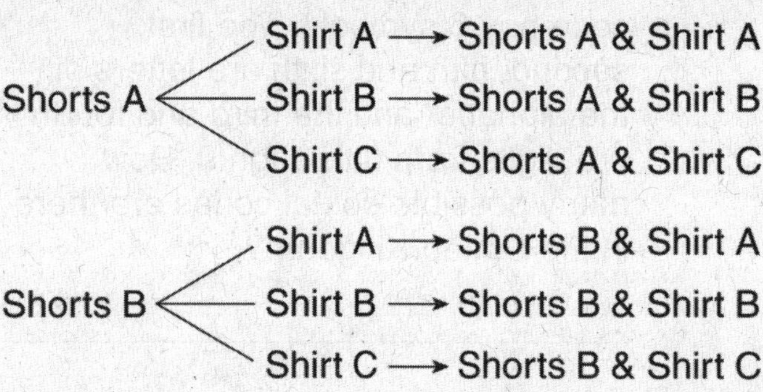

Use the tree diagram to answer the following questions.

1. How many different shirts could you pair with Shorts A?

____

2. How many different outfits could you make with Shorts A?

____

3. How many different shirts could you pair with Shorts B?

____

4. How many different outfits can you make with Shorts B?

____

5. With 2 different pairs of shorts and 3 different shirts, how many different outfits can you make?

____

      Holt McDougal Mathematics

LESSON 10-6

Puzzles, Twisters & Teasers

It's FUN-damental!

Determine if the number of possible outcomes is correct for each situation below. Circle the letter next to your answer. Use the letters to solve the riddle.

1. 3 types of birds and
   2 types of cages
   possible outcomes: 6

   I correct   **A** incorrect

2. 5 destinations and
   3 modes of transportation
   possible outcomes: 8

   X correct   **W** incorrect

3. 4 colors and 3 sizes
   possible outcomes: 7

   M correct   **T** incorrect

4. 5 shirts, 3 pants, and 2 jackets
   possible outcomes: 30

   A correct   **Z** incorrect

5. 3 types of bagels and
   3 types of spreads
   possible outcomes: 9

   J correct   **K** incorrect

6. 4 main dishes, 5 appetizers,
   and 5 desserts
   possible outcomes: 14

   L correct   **V** incorrect

7. 3 destinations and
   4 months
   possible outcomes: 34

   P correct   **U** incorrect

8. 3 types of paint and
   1 type of paper
   possible outcomes: 3

   E correct   **B** incorrect

9. 3 types of soup and
   5 types of sandwiches
   possible outcomes: 15

   S correct   **U** incorrect

10. 10 type fonts, 5 type sizes,
    and 3 paper sizes
    possible outcomes: 18

    V correct   **D** incorrect

What did the wave say to the beach?

Nothing, ____ ____ ____ ____ ____ ____ **T** ____ ____ ____ ____ ____

      Holt McDougal Mathematics

LESSON
10-7

Practice A
Permutations and Combinations

Express each expression as a product of factors.

1. $6!$

2. $3!$

3. $7!$

4. $\dfrac{8!}{5!}$

5. $\dfrac{4!}{2!}$

6. $\dfrac{9!}{6!}$

Evaluate each expression.

7. $5!$

8. $9!$

9. $3!$

10. $8!$

11. $\dfrac{7!}{4!}$

12. $\dfrac{8!}{7!}$

13. $\dfrac{5!}{2!}$

14. $7! - 5!$

15. $(6 - 3)!$

16. $\dfrac{4!}{(6 - 2)!}$

17. $\dfrac{9!}{(8 - 3)!}$

18. $\dfrac{7!}{(9 - 4)!}$

19. An anagram is a rearrangement of the letters of a word or words to make other words. How many possible arrangements of the letters W, O, R, D, and S can be made? ___________

20. Janell is having a group of friends over for dinner and is setting the name cards on the table. She has invited 5 of her friends for dinner. How many different seating arrangements are possible for Janell and her friends at the table? ___________

21. How many different selections of 4 books can be made from a bookcase displaying 12 books? ___________

   Holt McDougal Mathematics

Name _________________________ Date _____________ Class _____________

Practice B
Permutations and Combinations

Evaluate each expression.

1. 10!

2. 13!

3. 11! − 8!

4. 12! − 9!

5. $\dfrac{15!}{8!}$

6. $\dfrac{18!}{12!}$

7. $\dfrac{13!}{(17-12)!}$

8. $\dfrac{19!}{(15-2)!}$

9. $\dfrac{15!}{(18-10)!}$

10. Signaling is a means of communication through signals or objects. During the time of the American Revolution, the colonists used combinations of a barrel, basket, and a flag placed in different positions atop a post. How many different signals could be sent by using 3 flags, one above the other on a pole, if 8 different flags were available?

11. From a class of 25 students, how many different ways can 4 students be selected to serve in a mock trial as the judge, defending attorney, prosecuting attorney, and the defendant?

12. How many different 4 people committees can be formed from a group of 15 people?

13. The girls' basketball team has 12 players. If the coach chooses 5 girls to play at a time, how many different teams can be formed?

14. A photographer has 50 pictures to be placed in an album. How many combinations will the photographer have to choose from if there will be 6 pictures placed on the first page?

Holt McDougal Mathematics

Practice C

LESSON 10-7

Permutations and Combinations

Evaluate each expression.

1. $\dfrac{16!}{(15-4)!}$

2. $\dfrac{21!}{(19-3)!}$

3. $\dfrac{17!}{5!(17-5)!}$

4. $_7P_3$

5. $_9P_4$

6. $_{10}P_8$

7. $_{18}P_2$

8. $_9C_2$

9. $_{11}C_5$

10. $_{13}C_{11}$

11. $_{15}C_3$

12. The music class has 20 students and the teacher wants them to practice in groups of 5. How many different ways can the first group of 5 be chosen?

13. Math, science, English, history, health, and physical education are the subjects on Jamar's schedule for next year. Each subject is taught in each of the 6 periods of the day. From how many different schedules will Jamar be able to choose?

14. The Hamburger Trolley has 25 different toppings available for their hamburgers. They have a $3 special that is a hamburger with your choice of 5 different toppings. Assume no toppings are used more than once. How many different choices are available for the special?

15. Many over the counter stocks are traded through Nasdaq, an acronym for the National Association of Securities Dealers Automatic Quotations. Most of the stocks listed on the Nasdaq use a 4-digit alphabetical code. For example, the code for Microsoft is MSFT. How many different 4-digit alphabetical codes could be available for use by the association? Assume letters cannot be reused.

Holt McDougal Mathematics

LESSON 10-7	**Review for Mastery**
	Permutations and Combinations

Factorial: a string of factors that counts down to 1

$$6! = 6 \cdot 5 \cdot 4 \cdot 3 \cdot 2 \cdot 1$$

To evaluate an expression with factorials, cancel common factors.

$$\frac{5!}{3!} = \frac{5 \cdot 4 \cdot \cancel{3} \cdot \cancel{2} \cdot \cancel{1}}{\cancel{3} \cdot \cancel{2} \cdot \cancel{1}} = 5 \cdot 4 = 20$$

Complete to evaluate each expression.

1. $\dfrac{7!}{4!} = \dfrac{7 \cdot 6 \cdot 5 \cdot \; \cdot \; \cdot \; \cdot}{\; \cdot \; \cdot \; \cdot}$

   $= 7 \cdot$ ______ $=$ ______

2. $\dfrac{6!}{(5-2)!} = \dfrac{6!}{\underline{\;\;}!} = \dfrac{\; \cdot \; \cdot \; \cdot \; \cdot \; \cdot}{\; \cdot \;}$

   $=$ ______ $=$ ______

Permutation: an arrangement in which order is important

> *wxyz* is not the same as *yxzw*

Apply the Fundamental Counting Principle to find how many permutations are possible using all 4 letters *w, x, y, z* with no repetition.

$$\underset{\text{1st letter}}{\underline{\quad 4 \quad}} \times \underset{\text{2nd letter}}{\underline{\quad 3 \quad}} \times \underset{\text{3rd letter}}{\underline{\quad 2 \quad}} \times \underset{\text{4th letter}}{\underline{\quad 1 \quad}} = 4! = 24 \text{ possible arrangements}$$

When you arrange *n* things, *n*! permutations are possible.

Complete to find the number of permutations.

3. In how many ways can 6 people be seated on a bench that seats 6?

   $6! = \dfrac{\; \cdot \; \cdot \; \cdot \; \cdot \; \cdot}{}$

   $=$ ______ possibilities

4. How many 5-digit numbers can be made using the digits 7, 4, 2, 1, 8 without repetitions?

   $\underline{\;\;}! = \dfrac{\; \cdot \; \cdot \; \cdot \; \cdot}{}$

   $=$ ______ possibilities

Apply the Fundamental Counting Principle to find how many permutations are possible using 4 letters 2 at a time, with no repetitions.

$$\underset{\text{1st letter}}{\underline{\quad 4 \quad}} \times \underset{\text{2nd letter}}{\underline{\quad 3 \quad}} = 12 \text{ possible 2-letter arrangements}$$

Apply the Fundamental Counting Principle.

5. In how many ways can 6 people be seated on a bench that seats 4?

   $\underset{\text{1st seat}}{\underline{\quad\quad\quad}} \times \underset{\text{2nd seat}}{\underline{\quad\quad\quad}} \times \underset{\text{3rd seat}}{\underline{\quad\quad\quad}} \times \underset{\text{4th seat}}{\underline{\quad\quad\quad}} =$ ______ possibilities

6. How many 3-digit numbers can be made using the digits 7, 4, 2, 1, 8 without repetitions?

   $\underset{\text{1st digit}}{\underline{\quad\quad\quad}} \times \underset{\text{2nd digit}}{\underline{\quad\quad\quad}} \times \underset{\text{3rd digit}}{\underline{\quad\quad\quad}} =$ ______ possibilities

Holt McDougal Mathematics

<table><tr><td>LESSON
10-7</td><td>

Review for Mastery
Permutations and Combinations (continued)
</td></tr></table>

When using fewer than the available number of items in an arrangement, instead of the Fundamental Counting Principle, you can use a formula to find the number of possible permutations.

To arrange *n* things *r* at a time, the number of possible permutations *P* is: $_nP_r = \dfrac{n!}{(n-r)!}$

Find how many permutations are possible using 4 letters 2 at a time, with no repetitions.

$$_4P_2 = \frac{4!}{(4-2)!} = \frac{4!}{2!} = \frac{4 \cdot 3 \cdot \cancel{2} \cdot \cancel{1}}{\cancel{2} \cdot \cancel{1}} = 12 \text{ possible 2-letter arrangements}$$

Complete to apply the permutations formula.

7. In how many ways can 6 people be seated on a bench that seats 4?

$$_6P_4 = \frac{6!}{(6-\;\;)!} = \frac{\;\;!}{\;\;!} = \underline{\hspace{3cm}} = \underline{\hspace{1.5cm}} \text{ possible seating arrangements}$$

8. How many 3-digit numbers can be made using the digits 7, 4, 2, 1, 8 without repetitions?

$$_5P_{\underline{\;\;}} = \frac{5!}{(5-\;\;)!} = \frac{\;\;!}{\;\;!} = \underline{\hspace{3cm}} = \underline{\hspace{1.5cm}} \text{ possible 3-digit numbers}$$

Combination: an arrangement in which order is not important

How many 2-letter combinations can be made from the 4 letters *w*, *x*, *y*, *z* without repetition?

The combinations *w x* and *x w* are the same. After all the same combinations are removed, there are 6 different combinations possible.

The number of combinations *C* of *n* things taken *r* at a time is:

$$_4C_2 = \frac{_4P_2}{2!} = \frac{4!}{(4-2)!\,2!} = \frac{4!}{2!\,2!} = \frac{\overset{2}{\cancel{4}} \cdot 3 \cdot \cancel{2} \cdot \cancel{1}}{\underset{1}{\cancel{2}} \cdot 1 \cdot \cancel{2} \cdot \cancel{1}} = 2 \cdot 3 = 6$$

wx   x̶w̶   y̶w̶   z̶w̶
wy   xy   y̶x̶   z̶x̶
wz   xz   yz   z̶y̶

There are fewer combinations than permutations

$$_nC_r = \frac{_nP_r}{r!} = \frac{n!}{(n-r)!\,r!}$$

Complete to apply the combinations formula.

9. How many different 4-person committees can be formed from a group of 6 people?

$$_6C_4 = \frac{_6P_4}{4!} = \frac{6!}{(6-\;\;)!\;\;!} = \frac{6!}{\;\;!\;\;!} = \underline{\hspace{3cm}} = \underline{\hspace{1.5cm}} \text{ possible 4-person committees}$$

     Holt McDougal Mathematics

Challenge

Roundtable Discussion

The number of ways in which 4 people can be seated *in a row*, on a
bench that seats 4 is $_4P_4$, or 4!.

$$\frac{4}{\text{1st seat}} \times \frac{3}{\text{2nd seat}} \times \frac{2}{\text{3rd seat}} \times \frac{1}{\text{4th seat}} = 4! = {_4P_4} = 24 \text{ different arrangements}$$

Now consider what happens if 4 people are seated *in a circle*,
around a round table that seats 4.

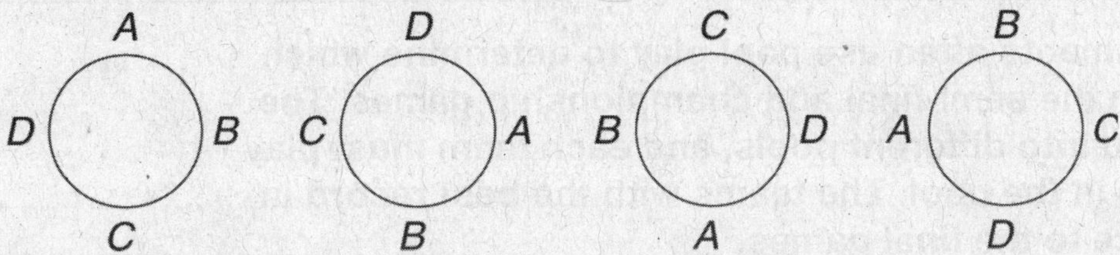

Note that the 4 circular arrangements shown are really all the same
with respect to who sits next to whom.

For each of the 4! permutations, there are 4 alike.

So, there are fewer ways to seat 4 people at a circular table that
seats 4.

$$\frac{_4P_4}{4} = \frac{4!}{4} = \frac{4 \cdot 3 \cdot 2 \cdot 1}{4} = 6 \text{ different arrangements}$$

1. In how many different ways can 5 people be seated in a row, on
   a bench that seats 5?

   ____

2. In how many different ways can 5 people be seated in a circle,
   around a circular table that seats 5?

   ____

3. In how many different ways can *n* people be seated in a row, on
   a bench that seats *n*? Answer in factorial form.

   ____

4. In how many different ways can *n* people be seated in a circle,
   around a circular table that seats *n*? Answer in factorial form.

   ____

Name _________________________________ Date ________________ Class ______________

Problem Solving

Permutations and Combinations

Write the correct answer.

1. In a day camp, 6 children are picked to be team captains from the group of children numbered 1 through 49. How many possibilities are there for who could be the 6 captains?

2. If you had to match 6 players in the correct order for most popular outfielder from a pool of professional players numbered 1 through 49, how many possibilities are there?

Volleyball tournaments often use pool play to determine which teams will play in the semi-final and championship games. The teams are divided into different pools, and each team must play every other team in the pool. The teams with the best record in pool play advance to the final games.

3. If 12 teams are divided into 2 pools, how many games will be played in each pool?

4. If 12 teams are divided into 3 pools, how many pool play games will be played in each pool?

A word jumble game gives you a certain number of letters that you must make into a word. Choose the letter for the best answer.

5. How many possibilities are there for a jumble with 4 letters?

   A 4   C 24

   B 12   D 30

6. How many possibilities are there for a jumble with 5 letters?

   F 24   H 120

   G 75   J 150

7. How many possibilities are there for a jumble with 6 letters?

   A 120

   B 500

   C 720

   D 1000

8. On the Internet, a site offers a program that will un-jumble letters and give you all of the possible words that can be made with those letters. However, the program will not allow you to enter more than 7 letters due to the amount of time it would take to analyze. How many more possibilities are there with 8 letters than with 7?

   F 5040   H 35,280

   G 20,640   J 40,320

Holt McDougal Mathematics

Reading Strategies

Use a Visual Aid

A **permutation** is an arrangement of objects in a certain order.

How many different ways can you arrange these three shapes?

You can use a tree diagram to visualize all of the possible arrangements:

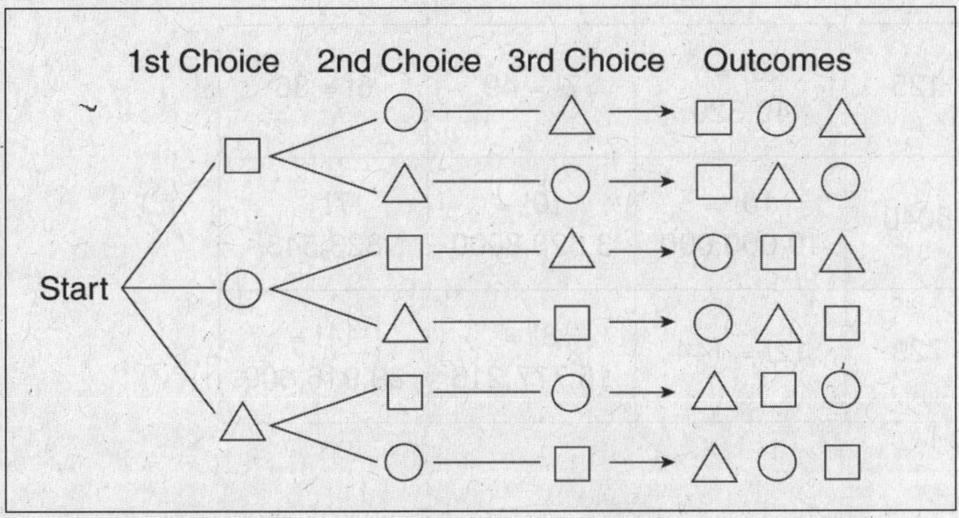

Use the tree diagram to answer the following.

1. If you start with the circle, how many different arrangements can you make? List them.

____

2. If you start with the square, how many different arrangements can you make? List them.

____

3. If you start with the triangle, how many different arrangements can you make? List them.

____

4. How many different arrangements can you make with these three shapes?

____

Holt McDougal Mathematics

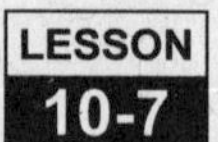

LESSON 10-7
Puzzles, Twisters & Teasers
Finding a Treasure!

Black out the incorrect expressions to see a shape.

9! = 362,880	2! = 4	11! = 9,497,876	3! = 9	5! = 120
5! = 25	3! = 6	4! = 14	2! = 2	10! = 100
6! = 150	5! = 125	8! = 40,320	7! = 49	6! = 36
4! = 256	7! = 5040	10! = 10,000,000	10! = 3,628,8000	7! = 823,543
4! = 24	9! = 729	12! = 144	8! = 16,777,216	11! = 39,916,800

What do you see? ______________________________

Holt McDougal Mathematics

Answers

LESSON 10-1

Practice A

1. 0.7 or 70%; 0.3 or 30%

2. $\dfrac{1}{4}$

3. $\dfrac{1}{4}$

4. $\dfrac{1}{2}$

5. 0

6. $\dfrac{1}{2}$

7. 1

8. $\dfrac{1}{2}$

9. $\dfrac{3}{4}$

10. 1

11. $P(\text{boy}) = \dfrac{14}{25}$; $P(\text{girl}) = \dfrac{11}{25}$

12. $\dfrac{1}{2}$

Practice B

1. $\dfrac{3}{32}$

2. $\dfrac{1}{16}$

3. $\dfrac{3}{8}$

4. $\dfrac{3}{4}$

5. $\dfrac{3}{8}$

6. 0

7. $\dfrac{1}{4}$

8. $\dfrac{5}{8}$

9. 0.035, or 3.5%

10. 0.175, or 17.5%

11. 0.628, or 62.8%

12.

Outcome	Jordan	Stacey	Diane	Carlos
Probability	45%	25%	20%	10%

Practice C

1. $\dfrac{7}{25}$

2. $\dfrac{4}{25}$

3. $\dfrac{7}{20}$

4. $\dfrac{21}{25}$

5. $\dfrac{2}{25}$

6. $\dfrac{63}{100}$

7. $\dfrac{1}{8}$

8. $\dfrac{1}{4}$

9. $\dfrac{1}{8}$

10. $\dfrac{1}{8}$

11. $\dfrac{3}{8}$

12. $\dfrac{3}{8}$

13. $\dfrac{5}{8}$

14. $\dfrac{3}{4}$

15.

Outcome	Probability
Team A	0.15
Team B	0.3
Team C	0.15
Team D	0.15
Team E	0.15
Team F	0.1

Review for Mastery

1. heads, tails; 2; $\dfrac{1}{2}$

2. 1, 2, 3, 4, 5, 6; 6; $\dfrac{1}{6}$

3. 50%

4. 0.28

5. 40%

6. 0.65

7. $P(A)$; 8; 7; 15; $\dfrac{1}{2}$; $\dfrac{1}{2}$; 50

8. $P(F)$; 2; 1; 3; $\dfrac{1}{10}$; $\dfrac{1}{10}$; 10

Holt McDougal Mathematics

Challenge

1. TT
2. T T
3. a. b.

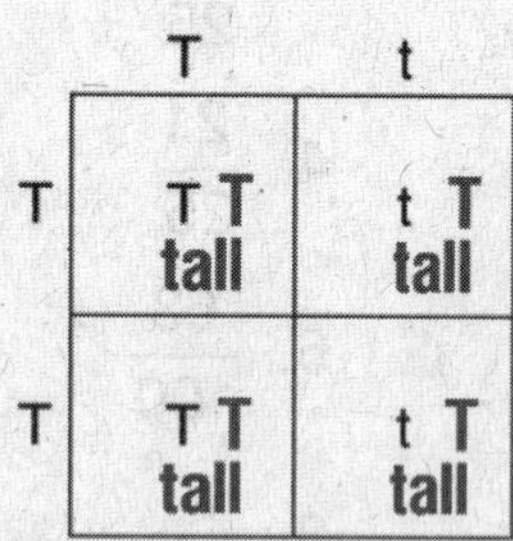

c. $\frac{4}{4}$ or 1 or 100%

4. a.

b. $\frac{3}{4}$ or 75%

Problem Solving

1. $\frac{3}{4}$
2. $\frac{13,983,815}{13,983,816}$
3. 0.027, or 2.7%
4. 0.973, or 97.3%   5. B
6. F   7. C

Reading Strategies

1. probability   2. P(event)
3. 1 or 100%   4. an experiment
5. a trial
6. the result of a trial

Puzzles, Twisters & Teasers

Across
3. Complement
5. One
7. Event
8. Trial
Down
1. Outcome
2. Probability
4. Experiment
6. Sample

LESSON 10-2

Practice A

1. $\frac{3}{100}$ or 3%   2. $\frac{1}{25}$ or 4%
3. $\frac{19}{50}$ or 38%   4. $\frac{1}{100}$ or 1%
5. $\frac{21}{50}$ or 42%   6. $\frac{3}{25}$ or 12%
7. 0.0625; 0.375; less likely
8. 0.1875; 0.125; more likely
9. $\frac{4}{25}$ or 16%

Practice B

1. $\frac{1}{4}$ or 25%   2. 60%
3. about 17%   4. Guy
5. about 27%

Practice C

1. $\frac{1}{8}$ or 12.5%   2. ≈0.1212
3. ≈0.0303   4. ≈0.0606
5. ≈0.0909   6. ≈0.1515
7. 0
8. ≈0.0606; ≈0.1212; less likely
9. about 20%

Holt McDougal Mathematics

Review for Mastery

1. 50; 1
2. 100; 1
3. **Experimental Probability** (ratio)

$40, \dfrac{1}{10}; 100, \dfrac{1}{4}; 80, \dfrac{1}{5}; 60, \dfrac{3}{20};$

$120, \dfrac{3}{10}$

Experimental Probability (percent)
10%; 25%; 20%; 15%; 30%

Challenge

1. Results will vary.
2. $\dfrac{1}{2}; \dfrac{1}{2}$
3. near $\dfrac{1}{2}$; near $\dfrac{1}{2}$
4. Possible answer: very close
5. Results will vary.
6. $\dfrac{1}{2}; \dfrac{1}{2}$
7. Answers will vary.
8. Results will vary.
9. $\dfrac{1}{6}; \dfrac{1}{6}; \dfrac{1}{6}; \dfrac{1}{6}; \dfrac{1}{6}; \dfrac{1}{6}$
10. vary; vary; vary; vary; vary; vary
11. Possible answer: very close

Problem Solving

1. 48%
2. 59%
3. 70%
4. 42%
5. 30%
6. 63%
7. C
8. G
9. B
10. G

Reading Strategies

1. the striped section, because it is the largest section
2. the dotted section, because it is the smallest section
3. 163

4. No; The striped section is the largest, but the spinner landed more times on the white section.
5. Yes; The spotted section is the smallest and the spinner landed on it the least number of times.

Puzzles, Twisters & Teasers

1. 20.1%
2. 28.5%
3. 13%
4. 23.3%
5. 17%
6. 35%
7. 18%
8. 11%
9. 3.5%
10. 21%

H A V E S A N D Y C L A W S

LESSON 10-3

Practice A

1. HH; HT; TH; TT
2. $\dfrac{1}{2}$
3. $\dfrac{1}{2}$
4. $\dfrac{1}{6}$
5. $\dfrac{1}{6}$
6. $\dfrac{1}{2}$
7. $\dfrac{1}{3}$
8. $\dfrac{1}{36}$
9. $\dfrac{1}{12}$
10. $\dfrac{1}{36}$
11. $\dfrac{1}{12}$
12. $\dfrac{1}{36}$
13. $\dfrac{1}{4}$
14. 8.72%
15. $\dfrac{1}{2}$

Practice B

1. $\dfrac{1}{6}$
2. 0
3. $\dfrac{1}{3}$
4. $\dfrac{5}{6}$

Holt McDougal Mathematics

5. $\dfrac{2}{3}$　　　　6. $\dfrac{1}{3}$

7. $\dfrac{2}{3}$　　　　8. $\dfrac{1}{2}$

9. $\dfrac{1}{18}$　　　10. $\dfrac{1}{6}$

11. $\dfrac{1}{9}$　　　12. $\dfrac{1}{36}$

13. $\dfrac{1}{12}$　　14. 0

15. $\dfrac{5}{18}$　　16. 1

17. $\dfrac{5}{12}$　　18. 40%

19. $\dfrac{4}{9}$

Practice C

1. $\dfrac{1}{9}$　　　2. $\dfrac{11}{12}$

3. $\dfrac{1}{12}$　　4. $\dfrac{35}{36}$

5. $\dfrac{7}{12}$　　6. $\dfrac{1}{6}$

7. $\dfrac{1}{8}$　　　8. $\dfrac{1}{8}$

9. $\dfrac{1}{4}$　　10. $\dfrac{1}{2}$

11. 1　　12. 6.69%

13. $\dfrac{5}{12}$

Review for Mastery

1. $\dfrac{1}{5}$; $\dfrac{2}{5}$　　2. $\dfrac{2}{4}$; $\dfrac{1}{2}$; $\dfrac{1}{4}$

3. $\dfrac{2}{6}$; $\dfrac{1}{3}$; $\dfrac{4}{6}$; $\dfrac{2}{3}$　　4. about 35.71%

5. 7, 9;
   −6, −4;
   yes; no number in both

6. −6, 9;
   −6, −4, 0, 2, 4;
   no; −6 in both events

7. −6, −4
   3
   2
   3; 2; 5

8. 7, 9
   −6, −4, 0, 6
   $\dfrac{2}{7}$
   $\dfrac{4}{7}$
   $\dfrac{2}{7} + \dfrac{4}{7} = \dfrac{6}{7}$

Challenge

1. odd; >2;

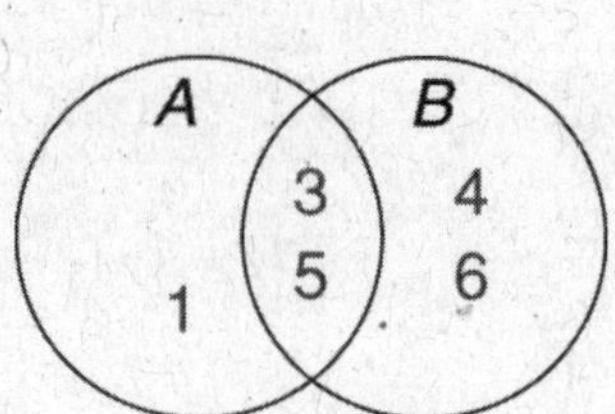

$\dfrac{2}{6}$, or $\dfrac{1}{3}$

2. even; >3

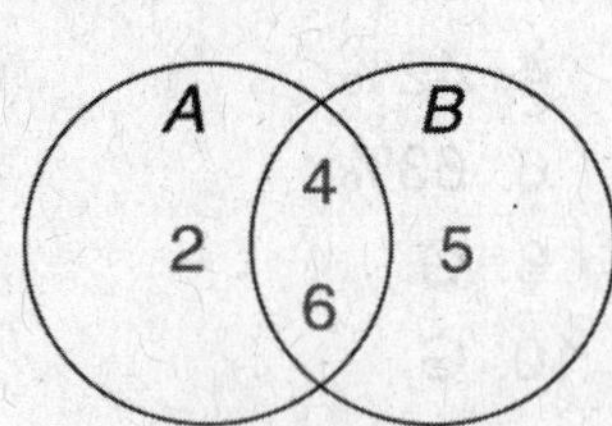

$\dfrac{4}{6}$, or $\dfrac{2}{3}$

Holt McDougal Mathematics

Problem Solving

1. $\dfrac{1}{100,000}$; 0.001%

2. $\dfrac{1}{200}$; 0.5%   3. $\dfrac{3}{100}$; 3%

4. $\dfrac{3,501}{100,000}$; 3.501%

5. B

	D	d
D	DD	Dd
D	Dd	dd

6. G   7. B

8. H

Reading Strategies

1. 6   2. 2 ways

3. $\dfrac{2}{6}$ or $\dfrac{1}{3}$ 4. 4 ways

5. $\dfrac{4}{6}$ or $\dfrac{2}{3}$

6. No: The probability of rolling a 2 is $\dfrac{2}{6}$ and the probability of rolling of an even number is $\dfrac{4}{6}$.

7. You are more likely to roll an even number than an odd number.

Puzzles, Twisters & Teasers

E: $\dfrac{3}{8}$; O: $\dfrac{3}{8}$; P: $\dfrac{1}{8}$; F: $\dfrac{3}{8}$; A: $\dfrac{1}{8}$; S: $\dfrac{3}{8}$; R: $\dfrac{3}{8}$; K: $\dfrac{3}{8}$; C: 0; Y: 0

O F F A
S K Y S C R A P E R

LESSON 10-4

Practice A

1. independent   2. dependent
3. independent   4. dependent

5. $\dfrac{1}{12} = 0.08\overline{3}$ 6. $\dfrac{1}{8} = 0.125$

7. $\dfrac{1}{24} = 0.0.041\overline{6}$ 8. $\dfrac{1}{6} = 0.1\overline{6}$

9. $\dfrac{5}{24} = 0.208\overline{3}$ 10. $\dfrac{1}{8} = 0.125$

11. $\dfrac{1}{8} = 0.125$

Practice B

1. independent   2. independent

3. dependent   4. $\dfrac{1}{12}$

5. $\dfrac{1}{4}$ 6. $\dfrac{1}{6}$

7. $\dfrac{1}{5} = 0.2$ 8. $\dfrac{1}{15} = 0.0\overline{6}$

9. $\dfrac{1}{30} = 0.0\overline{3}$ 10. $\dfrac{1}{40} = 0.025$

11. $\dfrac{1}{20} = 0.05$ 12. $\dfrac{1}{8} = 0.125$

13. $\dfrac{1}{120} = 0.008\overline{3}$ 14. $\dfrac{1}{24} = 0.041\overline{6}$

15. 0

Holt McDougal Mathematics

Practice C

1. $\dfrac{1}{676} \approx 0.00148$

2. $\dfrac{1}{104} \approx 0.00962$

3. $\dfrac{1}{2652} \approx 0.000377$

4. $\dfrac{4}{663} \approx 0.00603$

5. $\dfrac{13}{51} \approx 0.2549$

6. $\dfrac{1}{34} \approx 0.0294$

7. $\dfrac{1}{2} = 0.5$

8. $\dfrac{2}{21} \approx 0.0952$

9. $\dfrac{1}{15,600} \approx 0.000064$

10. $\dfrac{1}{12} \approx 0.08\overline{3}$

Review for Mastery

1. independent

2. dependent

3. independent

4. $2; 3; \dfrac{1}{24}$

5. $2; 3; \dfrac{1}{22}$

6. $2; \dfrac{2}{12}; \dfrac{1}{36}$

7. $2; \dfrac{1}{11}; \dfrac{1}{66}$

Challenge

1. $\dfrac{43}{50}$

2. $\dfrac{7}{50}$

3. $\dfrac{1}{5}$

4. $\dfrac{4}{25}$

5. $\dfrac{1}{10}$

Problem Solving

1. Independent events

2. Dependent events

3. game: $\dfrac{1}{13,983,816}$;

   Coin: $\dfrac{1}{16,777,216}$

4. Pick 6 game

5. B   6. H

7. A   8. G

Reading Strategies

1. independent events

2. independent events

3. dependent events

4. independent events

5. dependent events

Puzzles, Twisters & Teasers

1. I   2. R

3. E   4. U

5. B   6. S

7. C   8. G

9. E   10. R

I C E B U R G E R S

LESSON 10-5

Practice A

1. $\dfrac{3}{10}$

2. 450 tiger caps

3. $\dfrac{1}{4}$

4. 375 panda bear caps

5. 300 orangutan caps

6. 15 times

7. fair: $\dfrac{1}{4} = \dfrac{1}{4}$

Holt McDougal Mathematics

Practice B

1. 270

2. 225

3. 675

4. 5

5. not fair: $\dfrac{1}{6} \neq \dfrac{5}{6}$

6. fair: $\dfrac{5}{18} = \dfrac{5}{18}$

7. not fair: $\dfrac{3}{8} \neq \dfrac{5}{8}$

Practice C

1. 7 times

2. 25 times

3. 50 times

4. 15 times

5. 32 times

6. 30 times

7. 10 times

8. otter: 675 votes; osprey: 300 votes; beaver: 225 votes

9. fair: $\dfrac{1}{2} = \dfrac{1}{2}$

10. not fair: $\dfrac{5}{12} \neq \dfrac{5}{18}$

11. not fair: $\dfrac{3}{8} \neq \dfrac{1}{4}$

Review for Mastery

1. a. $\dfrac{1}{4}$

   b. $\dfrac{1}{4}$; 8

   c. 8

2. a. $\dfrac{1}{4}$

   b. $48 \times \dfrac{1}{4}$; 12

   c. 12

3. a. $\dfrac{1}{2}$

   b. $50 \times \dfrac{1}{2}$; 25

   c. 25

Challenge

1. a. $\dfrac{1}{12}$

   b. Possible answer: rolling a sum greater than 3

   c. $\dfrac{11}{12}$

2. a. $\dfrac{1}{6}$

   b. Possible answer: rolling a sum less than 10

   c. $\dfrac{5}{6}$

3. a. $\dfrac{5}{36}$

   b. Possible answer: rolling a sum other than 8

   c. $\dfrac{31}{36}$

Problem Solving

1. 150 defective bulbs

2. not fair; $\dfrac{1}{12} \neq \dfrac{1}{6}$

3. 16 times

4. 20 times

5. B

6. F

7. C

Reading Strategies

1. 0.4

2. $0.4 \times 2{,}000$

3. 800

4. 0.3

5. 600

Holt McDougal Mathematics

Puzzles, Twisters & Teasers

FRIENDSHIP

LESSON 10-6

Practice A

1. small tea, small juice, small milk; medium tea, medium juice, medium milk; large tea, large juice, large milk

2. purple jerseys with purple pants
purple jerseys with white pants
white jerseys with purple pants
white jerseys with white pants
striped jerseys with purple pants
striped jerseys with white pants

3. $\dfrac{1}{6}$

4. 26,000

5. $\dfrac{1}{26,000}$

6. about 46.67%

Practice B

1. 67,600

2. $\dfrac{1}{67,600} \approx 0.000015$

3. $\dfrac{65,000}{67,600} = \dfrac{25}{26} \approx 0.962$

4. $\dfrac{54,756}{67,600} = \dfrac{81}{100} = 0.81$

5.

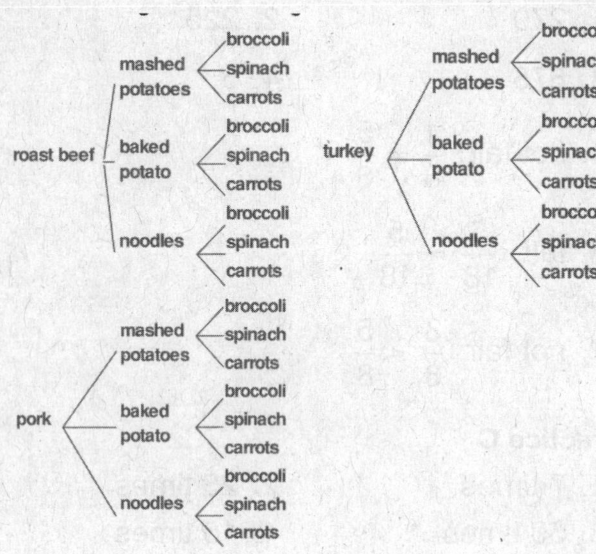

6. 27

7. $\dfrac{1}{3} = 0.33\overline{3}$

8. $\dfrac{1}{9} = 0.11\overline{1}$

9. about 46.67%

Practice C

1. 6

2. 12

3. 24

4. 84

5. 24

6. 108

7. 36

8. $\dfrac{9}{50} = 0.18$

9. $\dfrac{1875}{17,576} \approx 0.1067$

Review for Mastery

1. 6 × 4, 24

2. 2 × 4 × 3, 24

3.

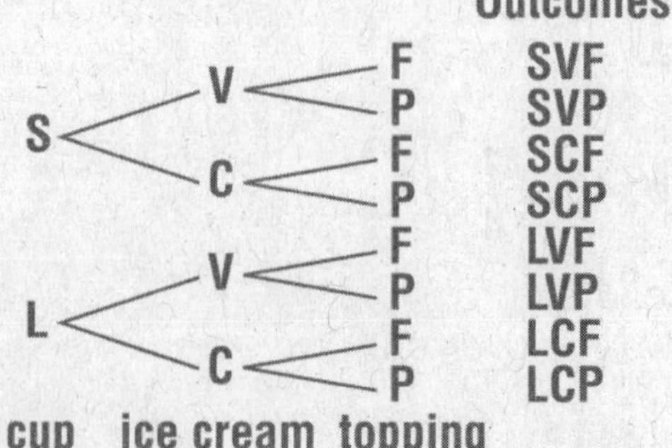

Holt McDougal Mathematics

4. a. 6; 5; 4; 3; 360

   b. $\dfrac{1}{360}$

   c. 5; 4; 3; 2; 120; $\dfrac{120}{360}$; $\dfrac{1}{3}$

Challenge

1. 8 × 10 × 10, or 800
2. 10 × 10 × 10 × 10, or 10,000
3. 800 × 10,000, or 8,000,000
4. 8,000,000 – 8 × 10 × 10 × 10 × 10, or 7,920,000
5. 10,000

Problem Solving

1. 100,000
2. 1,000,000,000
3. 17,576,000
4. 45,697,600
5. B
6. H
7. B
8. G

Reading Strategies

1. 3 different shirts
2. 3 different outfits
3. 3 different shirts
4. 3 different outfits
5. 6 different outfits

Puzzles, Twisters & Teasers

1. I
2. W
3. T
4. A
5. J
6. V
7. U
8. E
9. S
10. D

I T J U S T W A V E D

LESSON 10-7

Practice A

1. 6•5•4•3•2•1
2. 3•2•1
3. 7•6•5•4•3•2•1

4. $\dfrac{8 \cdot 7 \cdot 6 \cdot 5 \cdot 4 \cdot 3 \cdot 2 \cdot 1}{5 \cdot 4 \cdot 3 \cdot 2 \cdot 1}$

5. $\dfrac{4 \cdot 3 \cdot 2 \cdot 1}{2 \cdot 1}$

6. $\dfrac{9 \cdot 8 \cdot 7 \cdot 6 \cdot 5 \cdot 4 \cdot 3 \cdot 2 \cdot 1}{6 \cdot 5 \cdot 4 \cdot 3 \cdot 2 \cdot 1}$

7. 120
8. 362,880
9. 6
10. 40,320
11. 210
12. 8
13. 60
14. 4920
15. 6
16. 1
17. 3024
18. 42
19. 120
20. 720
21. 495

Practice B

1. 3,628,800
2. 6,227,020,800
3. 39,876,480
4. 478,638,720
5. 32,432,400
6. 13,366,080
7. 51,891,840
8. 19,535,040
9. 32,432,400
10. 336
11. 303,600
12. 1365
13. 792
14. 15,890,700

Practice C

1. 524,160
2. 2,441,880
3. 6188
4. 210
5. 3024
6. 1,814,400
7. 306
8. 36
9. 462
10. 78
11. 455
12. 15,504 groups
13. 720 schedules
14. 53,130 choices
15. 358,800 codes

Review for Mastery

1. $\dfrac{4 \cdot 3 \cdot 2 \cdot 1}{4 \cdot 3 \cdot 2 \cdot 1}$; 6•5; 210

2. 3; $\dfrac{6 \cdot 5 \cdot 4 \cdot 3 \cdot 2 \cdot 1}{3 \cdot 2 \cdot 1}$; 6•5•4; 120

3. 6•5•4•3•2•1; 720
4. 5; 5•4•3•2•1; 120

Holt McDougal Mathematics

5. 6; 5; 4; 3; 360

6. 5; 4; 3; 60

7. $4; \dfrac{6}{2}; \dfrac{6 \cdot 5 \cdot 4 \cdot 3 \cdot 2 \cdot 1}{2 \cdot 1}; 360$

8. $3; 3; \dfrac{5}{2}; \dfrac{5 \cdot 4 \cdot 3 \cdot 2 \cdot 1}{2 \cdot 1}; 60$

9. $4; 4; 2; 4; \dfrac{6 \cdot 5 \cdot 4 \cdot 3 \cdot 2 \cdot 1}{2 \cdot 1 \cdot 4 \cdot 3 \cdot 2 \cdot 1}; 15$

Challenge

1. $_5P_5 = 5! = 5 \cdot 4 \cdot 3 \cdot 2 \cdot 1 = 120$

2. $\dfrac{_5P_5}{5} = \dfrac{5!}{5} = \dfrac{5 \cdot 4 \cdot 3 \cdot 2 \cdot 1}{5} = 24$

3. $n!$

4. $(n-1)!$

Problem Solving

1. 13,983,816 possibilities

2. 10,068,347,520 possibilities

3. 15 games

4. 6 games

5. C

6. H

7. C

8. H

Reading Strategies

1. 2; circle, square, triangle; circle, triangle, square

2. 2; square, circle, triangle; square, triangle, circle,

3. 2; triangle, square, circle; triangle, circle, square

4. 6

Puzzles, Twisters & Teasers

9! = 362,880	2! = 4	11! = 9,497,876	3! = 9	5! = 120
5! = 25	3! = 6	4! = 14	2! = 2	10! = 100
6! = 150	5! = 125	8! = 40,320	7! = 49	6! = 36
4! = 256	7! = 5040	10! = 10,000,000	10! = 3,628,8000	7! = 823,543
4! = 24	9! = 729	12! = 144	8! = 16,777,216	11! = 39,916,800

the shape of a large X

Holt McDougal Mathematics